From Physics to Philosophy

This collection of essays by leading philosophers of physics offers philosophical perspectives on two of the central elements of modern physics, quantum theory and relativity. The topics examined include the notorious 'measurement problem' of quantum theory and the attempts to solve it by attributing extra values to physical quantities, the mysterious nonlocality of quantum theory, the curious properties of spatial localization in relativistic quantum theories, and the problem of time in the search for a theory of quantum gravity. Together the essays represent some of the most recent research in philosophy of physics, and break new ground within the philosophy of quantum theory.

JEREMY BUTTERFIELD is Senior Research Fellow at All Souls College, Oxford. He has published articles in analytical philosophy, especially philosophy of physics.
CONSTANTINE PAGONIS is a former Research Fellow at Wolfson College, Cambridge. He has published articles in the philosophy of physics.

From Physics to Philosophy

Edited by
JEREMY BUTTERFIELD and
CONSTANTINE PAGONIS

CAMBRIDGE
UNIVERSITY PRESS

CAMBRIDGE UNIVERSITY PRESS
Cambridge, New York, Melbourne, Madrid, Cape Town, Singapore,
São Paulo, Delhi, Dubai, Tokyo, Mexico City

Cambridge University Press
The Edinburgh Building, Cambridge CB2 8RU, UK

Published in the United States of America by Cambridge University Press, New York

www.cambridge.org
Information on this title: www.cambridge.org/9780521154475

First published 1999
First paperback printing 2010

A catalogue record for this publication is available from the British Library

Library of Congress Cataloguing in Publication data

From physics to philosophy / edited by Jeremy Butterfield and Constantine Pagonis.
p. cm.
Includes bibliographical references and index.
ISBN 0 521 66025 4 (hb)
1. Physics–Philosophy–Congresses. 2. Quantum theory–Congresses.
3. Relativity (Physics)–Congresses. I. Redhead, Michael. II. Butterfield,
Jeremy. III. Pagonis, Constantine.
QC5.56.F76 1999 99-11531
530′.01–dc21 CIP

ISBN 978-0-521-66025-9 Hardback
ISBN 978-0-521-15447-5 Paperback

Contents

Contributors

GORDON BELOT is Associate Professor of Philosophy, New York University

GARY BOWMAN is Assistant Professor of Physics, Northern Arizona University

HARVEY BROWN is Reader in Philosophy, and University Lecturer in Philosophy of Physics, Oxford University

JEREMY BUTTERFIELD is Senior Research Fellow, All Souls College, Oxford University

ROB CLIFTON is Associate Professor of Philosophy, University of Pittsburgh

JAMES CUSHING is Professor of Physics, University of Notre Dame

JOHN EARMAN is University Professor of History and Philosophy of Science, University of Pittsburgh

ARTHUR FINE is the John Evans Professor of Philosophy, Northwestern University

GORDON FLEMING is Professor of Physics (Emeritus), Pennsylvania State University

STEVEN FRENCH is Senior Lecturer, Division of History and Philosophy of Science, School of Philosophy, University of Leeds

CONSTANTINE PAGONIS is a former Research Fellow at Wolfson College, Cambridge

SIMON SAUNDERS is Lecturer in Philosophy of Science, Oxford University

ABNER SHIMONY is Professor Emeritus of Philosophy and Physics, Boston University

Preface

Most of the essays in this collection were presented as papers at a two-day conference, held in the History and Philosophy of Science Department, Cambridge University, in June 1997. The editors are very grateful to the Department, to the British Society for the Philosophy of Science, and to the Mind Association for their generous financial support of the conference.

The conference was held on the occasion of Michael Redhead's retirement from the Professorship of the History and Philosophy of Science in the University of Cambridge; and the authors and editors are delighted to honour him with this volume. A bibliography of his writings is included.

Introduction

JEREMY BUTTERFIELD AND CONSTANTINE PAGONIS

Physics is a vast discipline, rich with topics inviting philosophical analysis. In undertaking that analysis, the philosophy of physics can be pursued in different styles. There is a spectrum: ranging from the 'theorem-proof' style as used in expounding a piece of mathematical physics (though here the theorems will be motivated by philosophical ideas), through philosophical-cum-physical argument that adverts in detail to pieces of physics (e.g. in displayed equations), to purely philosophical argument that is in some way based on, or influenced by, physics. For the most part, the essays in this collection are examples of the second of these three styles: they aim to combine detailed presentation of some technicalities in theoretical physics with the dialectical rigour of analytic philosophy.

The essays also have, with one exception, a common subject-matter: the philosophy, or if you prefer foundations, of quantum theory. The first six essays fall squarely in that subject-matter. The seventh and eighth broaden the discussion: they address, respectively, philosophical aspects of (i) the search for a theory of quantum gravity, and (ii) quantum theory's use of group theory. The final essay, by Abner Shimony, is much more speculative: it assesses the conjecture that the fundamental laws of nature are products of evolution.

Arthur Fine's essay starts the volume, on a central topic in philosophy of quantum theory: proofs of quantum nonlocality. Fine analyses a proof by Lucien Hardy, which as Fine says, is a characteristic example of a new generation of proofs that apparently make less use of probabilities and thereby dispense with inequalities. However, Fine shows that this appearance is misleading. Probabilities are substantially involved in Hardy's proof; and once we see this, we see that inequalities are also involved. Furthermore, Fine brings out how the theorem depends on the 'product rule', i.e. an algebraic condition on the assignment of values to physical quantities, which is used as a premise in Kochen–Specker-style no-hidden-variable proofs – and whose role in nonlocality proofs Fine long ago exposed! So Fine concludes that this unnoticed assumption means that Hardy's proof, though 'interesting and significant', is not 'a demonstration that quantum mechanics is nonlocal'.

Rob Clifton's essay is about the assignment of values to quantities in quantum theory. Clifton follows John Bell in wanting to promote some observables to the status of 'beables' – for which the probability of the quantity being a particular value is the probability of observing that value. But Clifton allows, in accordance with recent so-called 'modal' interpretations of quantum theory, that which observables are beables could depend on the quantum state; (unlike Bohmian mechanics, where position is picked out once and for all as the beable). Working in the framework of C^*-algebras, Clifton proposes an algebraic characterization of those sets of (bounded) observables which can be beables: namely, that the set be closed under all continuous self-adjoint functions of its members. Clifton calls such sets 'Segalgebras' in honour of Segal. He goes on to discuss assignments of values to these beables (as linear functionals on them). The idea is then that the measurement statistics given by a quantum state should be representable as an average over the actual values of the beables. Clifton then shows (his theorem 7) that this representation will be possible provided the Segalgebra is what he calls quasicommutative (roughly: the quotient Segalgebra, defined by factoring out the beables which are assigned 0 by the state, is commutative). (This result generalizes, and transposes to the algebraic setting, previous work by Bub & Clifton.) Clifton goes on to investigate maximal Segalgebras (having as many beables as possible), and to discuss two possible motivations for requiring commutativity, not just quasicommutativity.

The main purpose of Brown's essay is to explore aspects of the Galilean and gauge covariance of non-relativistic quantum mechanics, and their implications for the interpretation of the theory. He reviews first the nature of such covariance in relation to the single-body Schrödinger equation (for *arbitrary* background fields in the case of Galilean transformations) and he assesses the claim that the Maxwell field can be 'generated' by the local gauge principle when applied to the relativistic (Dirac) equation. He then describes how the geometric phase associated with Schrödinger evolution is not invariant under time-dependent spatial translations of the coordinate system (and hence under Galilean transformations), and briefly analyses the significance of this result. The remaining sections of Brown's paper concern the objectivity of 'sharp values' of observables (in particular as they appear in certain versions of the modal interpretation), in the light of recent work done jointly with M. Suárez and G. Bacciagaluppi, and the theory of quantum reference frames due to Aharonov and Kauffher. Brown emphasizes that it is only relational properties that can be considered frame- or gauge-independent elements of reality.

The next two essays are about the versions of quantum theory developed by de Broglie and Bohm. The first, by Saunders, presses a problem about these theories' ability to resolve the quantum measurement problem, in the relativistic domain. It has been suggested that the theory can be formulated using classical (c-number or Grassmanian) field configurations as 'beables'; the idea is to use a connection between the support of the non-relativistic 1-particle state, and the classical field configuration for associated states of the bosonic field. However, Saunders argues that this connection, far from ensuring that the beables are well-localized in measurement situations, implies exactly the opposite: the beables too are likely to develop into superpositions. He concludes that, since the *raison d'être* of the beables is to select one component, rather than another, of superpositions in the measurement context, the de Broglie–Bohm theory is undermined. He then explores the remaining interpretative options for the theory. *Prima facie*, there are three options; but Saunders argues that only the third – involving a commitment to the Dirac negative energy sea, in the fermionic case – can work.

Cushing & Bowman's essay is about Bohm's non-relativistic version of the theory, where particle-position is the beable; so-called 'Bohmian mechanics'. But their concern is not the measurement problem but that of empirically distinguishing this theory from orthodox 'Copenhagen' quantum theory. They begin by reviewing Bohmian mechanics, and its remarkable empirical equivalence, so far, to orthodox quantum theory. This makes it natural to look for some new domain, where one of the theories is better equipped conceptually to make a prediction. In view of the controversies surrounding the definition of 'quantum chaos' in the orthodox theory, and Bohmian mechanics' having the trajectories that are the prerequisite of classical chaos, Cushing & Bowman suggest quantum chaos as such a domain. This seems an especially promising domain, since Bohmian mechanics has a 'classical limit' quite unlike those of the orthodox theory; viz. that the quantum potential be zero. Work on chaos in Bohmian mechanics has only recently begun; but Cushing & Bowman's review suggests some tentative conclusions. The most tantalizing is that, since there are classical chaotic systems that cannot be reached as the limit of a Bohmian mechanical system, the ubiquity of classical chaos suggests that perhaps

Bohmian mechanics cannot account for all the physical phenomena we encounter.

Fleming & Butterfield's essay is about localization and associated concepts in Lorentz-invariant quantum theory: a controversial topic, on account of the apparent violation of Lorentz covariance and the superluminal propagation of localized states. They first review Newton & Wigner's fundamental work, stressing the similarities with the Galilean case; and this prompts them to review the difficulties of defining position operators for the Klein–Gordon and Dirac equations. The difficulties prompt the question: is there any standpoint from which the strange properties of relativistic localization make physical sense? The second half of the essay answers 'Yes' to this question: the standpoint is given by previous work by Fleming. Here, the exposition begins with *classical* Lorentz-covariant position variables; more specifically, three different parametrizations of them. These are equivalent for a point-particle; but for a localizable property of an extended system (such as the centre of energy or of charge), only one parametrization, the hyperplane-dependent parametrization, is convenient. Of the three, it is also the only one that can be consistently quantized. This in effect establishes the Yes answer to the above question. The essay ends by discussing some recent criticisms of this standpoint, by Malament and by Saunders.

The next two essays broaden the discussion, going beyond the philosophy of quantum theory. Belot & Earman's essay is a case-study in the 'symbiotic relationship' between physics and philosophy. They describe how the principal issue in the philosophy of general relativity, viz. the debate between substantivalism and relationism, is bound up with the difficulties facing the development of a quantum theory of gravity. They first point out how ironic, indeed unfortunate, it is that the recent philosophical debate, especially about Einstein's famous 'hole argument', has been conducted apparently in ignorance of the fact that the same issues were being debated in the quantum gravity community. Then they describe how formulating general relativity as a gauge theory makes clear the connections between the substantivalism–relationism debate, and physical questions about the definition of observables in classical general relativity – and above all, the disturbing fact that all gauge-invariant quantities in the theory are constant in time. This is the classical 'source' of quantum gravity's so-called 'problem of time': Belot & Earman go on to briefly survey four different approaches to it.

Steven French first discusses presentations of physical theories in terms of function spaces, and in terms of classes of models. French advocates a version of the latter approach: a version which uses partial models. He then reviews Wigner's and Weyl's great work on group theory in quantum physics as a case study of his model-theoretic approach. He ends by discussing how the case study, and this approach, bear on 'structural realism', as advocated by authors such as Worrall.

Abner Shimony's essay closes the volume in a suitably speculative and metaphysical way. He assesses the conjecture that the fundamental laws of nature are products of evolution. To do so, he first notes that for this conjecture to be coherent, there had better not be a fundamental law of natural selection; and indeed, according to other work by Shimony, there isn't. He then goes on to sketch some aspects of what one might call the 'cosmogonies' of two philosophers – Peirce and Whitehead – and two philosophical physicists – Wheeler and Smolin. He concludes on a sceptical but sympathetic note. He cannot endorse the conjecture; but the appeal to natural selection has several merits – for example, it provides 'a humble acknowledgement of the pervasiveness of contingency in the world and the futility of aiming at a completely rational world picture', and 'a multitude of instances of attaining an understanding of the emergence of order out of disorder'.

To conclude: our title reflects our belief that physics offers philosophers a vast store of topics that merit philosophical analysis. Though the philosophy of physics is at present a vigorous branch of philosophy, so much remains to be done! So, in the spirit of encouraging further work, we offer these essays to the community of philosophers of science and philosophically inclined physicists.

1

Locality and the Hardy theorem

ARTHUR FINE

But this conclusion [nonlocality] needs careful discussion in order to clarify what is going on. (Redhead 1987, p. 3)

Within the foundations of physics in recent years, Bell's theorem has played the role of what Thomas Kuhn calls a 'paradigm': that is, an exemplary piece of work that others learn from, imitate and develop. Following a period of articulation and consolidation, the first generation of developments of the Bell theorem was initiated by Heywood and Redhead (1983). They produced a nonlocality result in the algebraic style of the Bell–Kochen–Specker theorem (Bell 1966; Kochen and Specker 1967), moving away from the probabilistic relations characteristic of the Bell theorems proper. More recently a second generation develops results by Peres (1990), Greenberger–Horne–Zeilinger (1990), and Hardy (1993). In addition to moving away from probabilities, this generation tries to dispense with the limiting inequalities of the Bell theorem to yield so-called 'Bell theorems without inequalities'. With respect to probabilities, however, Hardy is a half-way house. It requires no inequalities but the result contradicts quantum mechanics under certain locality assumptions only if the statistical predictions of quantum mechanics hold in at least one case.

I want to examine the Hardy theorem and its interpretation. Initially, I intend to ignore respects in which it dispenses with probabilities because I want to point out the interesting significance of the theorem in a probabilistic context. We will see that when probabilities are restored, so are inequalities. Then we will see what the theorem has to contribute on the topic of locality.

1. Then Hardy example

According to Hardy, almost all the entangled states for a pair of systems give rise to a simple sort of Bell theorem. For our purposes a generic version of this will do.[1] So suppose we have a pair of systems whose state spaces are two dimensional (one can think of spin-1/2 systems, for instance), system I whose state space has orthonormal basis α, σ and system II with orthonormal basis β, τ. Denote by LC(...) a non-degenerate linear combination (that is, one with non-zero coefficients) of the enclosed terms. Let Ψ be the state of the combined (I + II) system. We will suppose that

$$\Psi = \mathrm{LC}(\alpha \otimes \tau, \sigma \otimes \beta, \sigma \otimes \tau). \tag{1a}$$

Collecting the terms, first in τ and then in σ, we get

$$\Psi = \mathrm{LC}(\alpha' \otimes \tau, \sigma \otimes \beta) \tag{1b}$$

and

$$\Psi = \mathrm{LC}(\alpha \otimes \tau, \sigma \otimes \beta') \tag{1c}$$

respectively, where $\alpha' = \mathrm{LC}(\alpha, \sigma)$ and $\beta' = \mathrm{LC}(\beta, \tau)$. Adding (1b) and (1c) and dividing by 2 yields

$$\Psi = \mathrm{LC}(\alpha' \otimes \tau, \sigma \otimes \beta', \sigma \otimes \beta, \alpha \otimes \tau). \tag{1d}$$

We can now read off various probabilistic statements from the form of these different representations of the joint state. For that purpose, let $A = P_{[\alpha]}$, $B = P_{[\beta]}$, $A' = I - P_{[\alpha']}$ and $B' = I - P_{[\beta']}$. Note that the relation between σ and A is not the same as that between α' and A'. Since there is no $\alpha \otimes \beta$ term in eqn 1a the result of projecting Ψ onto the $\alpha \otimes \beta$-space is null. Hence, where $P^{\Psi}(.)$ is the quantum probability in state Ψ, and writing $P^{\Psi}(AB)$ for $P^{\Psi}(A = 1 \ \& \ B = 1)$ – and so forth –

$$P^{\Psi}(AB) = 0. \tag{2a}$$

From eqn 1b, the result of projecting Ψ orthogonally to α' in the I-space leaves system II in state β, hence

$$P^{\Psi}(B|A') = 1. \tag{2b}$$

Similarly, from eqn 1c,

[1] The presentation below draws on Hardy's original (1993) and on the variant in Goldstein (1994). To map Hardy's discussion (U_i, D_i) onto my set U_1, $U_2 \leftrightarrow A$, B and D_1, $D_2 \leftrightarrow A'$, B'. The U_i are the same for Goldstein and his W_1, W_2 correspond (respectively) to my $I - A'$, $I - B'$.

$$P^{\Psi}(A|B') = 1. \tag{2c}$$

From eqn 1d, since neither $< \alpha|\alpha' >$ nor $< \beta|\beta' >$ is zero, the result of projecting Ψ so as to be orthogonal both to α' in the I-space and to β' in the II-space is not null. Hence,

$$P^{\Psi}(A'B') \neq 0. \tag{2d}$$

In eqn 1a the three-termed linear combination has non-zero coefficients the square of whose norms sum to 1, so $| < \Psi|\alpha \otimes \tau > |^2 + | < \Psi|\sigma \otimes \beta > |^2 \leq 1$. That is, $P^{\Psi}(A) + P^{\Psi}(B) \leq 1$. From eqn 2b, $P^{\Psi}(A') \leq P^{\Psi}(B)$. Adding $P^{\Psi}(A)$ to both sides yields

$$P^{\Psi}(A) + P^{\Psi}(A') \leq 1. \tag{2e}$$

Similarly from eqn 2c, $P^{\Psi}(B') \leq P^{\Psi}(A)$, which yields

$$P^{\Psi}(B) + P^{\Psi}(B') \leq 1. \tag{2f}$$

Extracting from the 0 to 1 probabilities, but using entirely similar reasoning about the geometry of the state space, the first three inferences of this series – eqns 2a, b, c – are the same as those drawn by Hardy. That is, talking about the values of quantities rather than probabilities for observing these values, he says that in state Ψ, $AB = 0$, that $(A' = 1) \Rightarrow (B = 1)$, and that $(B' = 1) \Rightarrow (A = 1)$. Hardy then draws the probabilistic conclusion 2d and urges that since local realism would sanction talk about locally possessed values it would lead instead to the conclusion that $A'B' = 0$. Hence he infers that a single A', B' measurement confirming the statistical prediction 2d would contradict local realism. Retaining the probabilities, the crux of the Hardy result would be a contradiction among eqns 2a, b, c, d that arises in the context of local hidden variables.

2. Random variables

The Bell inequalities have a purely probabilistic content as conditions governing whether a given set of probability distributions can be represented as the distributions of random variables. Given four pair distributions P_{AB}, $P_{A'B}$, $P_{AB'}$, $P_{A'B'}$ (say, on 0 and 1) with compatible singles P_A, P_B, $P_{A'}$, $P_{B'}$ (that is where the same marginal distribution P_A comes from P_{AB} and $P_{AB'}$, and so on) we can ask whether these fit together as the marginals of some four-distribution $P_{AA'BB'}$. Equivalently, we can ask whether there are random variables A, A', B and B' all defined on some common space whose single and joint distributions match the given singles and joints. Writing

$P(AB)$ for $P_{AB}(1, 1)$ – and so forth – necessary and sufficient for this is the satisfaction of the generalized Bell inequalities:

$$-1 \leq P(A'B') + P(A'B) + P(AB') - P(AB) - P(A') - P(B') \leq 0. \quad \text{(GB)}$$

Interchanging first A with A', then B with B' and finally both together yields a total of eight inequalities, which constitute the required necessary and sufficient conditions (Fine 1982a, b).

The connection with local hidden variables (or 'local realism') is just that a typical EPR-type correlation experiment (Einstein, Podolsky & Rosen 1935) yields pair distributions and compatible singles for quantum observables as above, where A, A' are noncommuting observables defined in one wing of the experiment and B, B' in the other. A local hidden variables model provides a way of representing those observables as random variables over a common space. In particular, locality is what justifies saying – for example – that $A(x)$, the value of A at 'hidden state' x, is well defined without regard to other observables, their values or measurements. Similarly for the probabilities represented by the $P(.)$ distribution. Thus the random variables framework codifies the idea that in a given state Ψ there are determinate values for the observables that do not depend on distant measurements and that are distributed according to definite probabilistic laws. As I will suggest below (section 4), I believe that standard uses of this framework employ principles beyond locality. Still, it is a clean and perspicuous way of treating locality conditions, a way that disentangles the discussion from murky 'elements of reality' and potentially misleading counterfactual reasoning ('If instead of measuring A' we had measured A then ... and also if instead of measuring B' we had measured B then, ...'). If we accept the standard framework, we can say that the (GB) inequalities provide the necessary and sufficient conditions for the existence of a local hidden variables model for a 2-by-2 EPR experiment.

Typically such an experiment works with imperfect correlations; that is, with joint probabilities neither 0 nor 1. We can ask, however, what would happen in the case where some one-way correlations are strict. Suppose, for instance, that measuring A' determined the outcome at B and that measuring B' determined the A outcome. That is, suppose, that the conditional probabilities $P(B|A')$ and $P(A|B')$ were both 1, as in the Hardy example. The following theorem characterizes this situation.

Theorem. If $P(B|A') = P(A|B') = 1$, then the pair distributions P_{AB}, $P_{A'B}$, $P_{AB'}$, $P_{A'B'}$, (on 0 and 1) with compatible singles P_A, P_B, $P_{A'}$, $P_{B'}$ are the distributions of random variables A, A', B and B' on some common space iff the following four conditions hold:

(i) $P(A'B') \leq P(AB)$
(ii) $P(A' \text{ or } B') \leq P(A \text{ or } B)$
(iii) $[P(A) - P(AB)] + [P(A') - P(A'B')] \leq 1$
(iv) $[P(B) - P(AB)] + [P(B') - P(A'B')] \leq 1.$

The theorem follows from the corresponding theorem for (GB) if we use conditions equivalent to $P(B|A') = P(A|B') = 1$; namely, that $P(A'B) = P(A')$ and $P(AB') = P(B')$. Indeed (i) is equivalent to the right side of (GB) under these assumptions. The Bell inequality $P(AB) + P(A'B) + P(AB') - P(A'B') \leq P(A) + P(B)$ is equivalent to (ii), that is, to $P(A') + P(B') - P(A'B') \leq P(A) + P(B) - P(AB)$. For (iii) the equivalent inequality is $P(A) + P(B') - 1 \leq P(AB) \; +P(A'B') +P(AB') - P(A'B)$. Interchanging A with A' and B with B' here yields, finally, the Bell inequality equivalent to (iv). Nothing new corresponds to the remaining four (GB) inequalities, each of which already follows from the stated conditions on the pair distributions.

We can now impose more structure, in particular the anti-correlations, $P(AB) = 0$, in the Hardy example.

Corollary 1. If $P(AB) = 0$, then the necessary and sufficient conditions reduce to

(a) $P(A'B') = 0$
(b) $\max[P(A) + P(A'), P(B) + P(B')] \leq 1$

Given that $P(AB) = 0$ and $0 \leq P(A'B')$, condition (i) holds iff (a) does. Similarly (b) is equivalent to (iii) and (iv). Finally by virtue of (a), (ii) automatically holds iff $P(AB) = 0$ since the assumptions of strict one-way correlation, that $P(A'B = P(A')$ and $P(AB') = P(B')$, imply that $P(A') \leq P(B)$ and that $P(B') \leq P(A)$.

Corollary 2. If $P(AB) = 0$ and $\max[P(A) + P(A'), P(B) + P(B')] \leq 1$, then

$$P(A'B') = 0 \text{ or, equivalently, } P(A'B') \leq P(AB)$$

is both necessary and sufficient for a random variables representation of the given joints and singles with strict correlations, $P(B|A') = P(A|B') = 1$.

3. The Hardy theorem

If we identify the quantum joints in the Hardy example with these distributions of random variables, all the requirements of corollary 2 are satisfied. From eqn 2a, $P(AB) = 0$. From eqn 2b, c, $P(B|A') = P(A|B') = 1$. From eqn 2e, f, $\max[P(A) + P(A'), P(B) + P(B')] \leq 1$. The condition $P(A'B') = 0$ that fails, according to eqn 2d, then, is precisely the condition whose satisfaction is both necessary and sufficient for a local hidden variables model of these probabilities. Thus for the Hardy case the equation $P(A'B') = 0$ (or the inequality $P(A'B') \leq P(AB)$) plays exactly the same role as do the (GB) inequalities for EPR in general. Although strict one-way correlations greatly simplify the reasoning, like Bell, Hardy has put his finger on precisely the central condition that makes local hidden variables possible for the case at hand.

Also, like the Bell theorem, the 'Hardy theorem' can be characterized as having two parts. One is a demonstration of a necessary condition for a hidden variables model. The second is the production of a generic example where that condition fails quantum mechanically. We can take

$$\text{if } P(B|A') = P(A|B') = 1 \text{ then } P(A'B') \leq P(AB)$$

for part (1). Then part (2) consists of the eqns 2b, c satisfying the 'if' clause and eqns 2a, d violating the consequent. Alternatively, we can take

$$\text{if } P(B|A') = P(A|B') = 1 \text{ and } P(AB) = 0, \text{ then } P(A'B') = 0$$

for part (1). Then eqns 2a, b, c satisfy the 'if' clause and eqn 2d violates the consequent.

We might call the first version the Hardy theorem 'with inequalities' and the second the Hardy theorem 'without inequalities'. Logically speaking, they are equivalent. Both versions identify the joint probabilities $P(..)$ that govern random variables with the quantum joint probabilities $P^{\Psi}(..)$ of the associated quantum observables. Indeed, it is that identification that makes part (2) of the theorem possible.

We can isolate what makes these versions equivalent. It is simply the probabilistic identity

$$P(A'B') = P(A'B'AB) \tag{3}$$

that must hold if all four variables are simultaneously representable. From eqn 3 it obviously follows that $P(A'B') \leq P(AB)$ and also that $P(A'B') = 0$

if $P(AB) = 0$. To see why eqn 3 holds note that in a random variables representation $P(A'B')$ is the marginal of a four-distribution as follows,

$$P(A'B') = P(A'B'AB) + P(A'B'\bar{A}B) + P(A'B'A\bar{B}) + P(A'B'\overline{AB}) \quad (4)$$

(where the bar means that the variable underneath takes the value 0). Since $P(A|A') = P(A|B') = 1$ iff $P(\bar{A}B') = P(A'\bar{B}) = 0$, all but the first term of eqn 4 vanishes – to produce eqn 3.

4. Probabilities and locality

Fans of the Hardy theorem may not be happy with my presentation of that result, even though it highlights the importance of the contradictory condition picked out by Hardy. My version is probabilistic and it depends critically on probabilistic reasoning and inequalities and, moreover, on the identification of quantum joint probabilities with random variable joints. How much simpler just to state that if $(A' = 1) \Rightarrow (B = 1)$ and $(B' = 1) \Rightarrow (A = 1)$, then if $AB = 0$, so too $A'B' = 0$. This is simpler, to be sure, in terms of reasoning to the conclusion, but not in terms of interpreting the quantum theory. For, like my presentation, the quantum theory is also probabilistic and to move from those probabilities to statements about values of quantities requires imposing an interpretive structure. Since the 'simple' inference does not hold in the quantum mechanical Hardy example, clearly his interpretative structure goes beyond the usual reading of the quantum probabilities in terms of likely outcomes of measurements. Of course it is supposed to do just that since the interpretation is supposed to require locality, in order to contradict it. We shall see, however, that it does more.

We can see exactly what the reading in terms of possessed values does require; namely, the principle that where the quantum joint probability is zero, as in eqn 2a, the observables in question do not both take the 0-probability values; that is, that either $A \neq 1$ or $B \neq 1$ in the case of eqn 2a. While this may seem like a harmless and modest principle, it is not. Years ago I showed that this principle is equivalent to the general Kochen–Specker functional condition: $f(Q)(x) = f[Q(x)]$ (Fine 1974). The connection is easy to see, for in any state Ψ and for any observable Q, $P^{\Psi}(Q = q \,\&\, f(Q) \neq f(q)) = 0$. So, the principle implies that $Q(x) = q$ only if $f(Q)(x) = f(q)$; that is, that $f(Q)(x) = f[Q(x)]$. This general functional condition, in turn, is equivalent to the product and sum rules: that the value possessed by the product of two observables (or their sum) is just obtained by multiplying (respectively, adding) the values possessed by the individual observables.[2] Even apart from locality considerations these rules are already inconsistent with the quantum theory. So the seemingly modest interpretive principle that governs the 'simple' form of the Hardy theorem is not at all harmless; indeed it is inconsistent with the quantum theory. The inconsistency runs deep and derives from how the basic framework of random variables is used.

The principle

$$\text{where } P^{\Psi}(A = q \ \& \ B = r) = 0, \ \text{ then } A \neq q \text{ or } B \neq r \tag{5a}$$

is in fact equivalent to the assumption about the random variables framework used in the two probabilistic versions of the Hardy theorem in section 3, that the joint probabilities of that framework match the quantum joints for observables where the latter are defined.[3] If it is locality that concerns us, this is not an assumption we need to make. For we can use the locality automatically built into the random variables framework – that values and probabilities are determinate and independent of distant measurements – and still entirely avoid this assumption on joint probabilities. Here is how. In a given state Ψ, make the usual association for a hidden variables construction between quantum observables and random variables. For each single observable this returns the quantum probabilities as the distribution of the associated random variable. To get joint probabilities where they are defined quantum mechanically (that is, for commuting observables) *do not* go to the (well-defined) joint distributions of the associated random variables. That would just re-instate the above principle. Instead, use the quantum mechanical identity

$$P^{\Psi}(A = q \ \& \ B = r) = P^{\Psi}(\chi_q(A)\chi_r(B) = 1) \tag{5b}$$

where $\chi(.)$ is the characteristic function (that is, $\chi_q(x) = 1$ for $x = q$ and 0 otherwise). The product observable $\chi_q(A)\chi_r(B)$ will correspond to some random variable, say C, whose distribution is quantum mechanical. If we assign the quantum joint probabilities by identifying $P(C = 1)$ with the right

[2] The equivalence also requires the rule ('spectrum rule') that the only possible values assigned to an observable in a state are those with non-zero probability in that state. I assume this in the discussions below.

[3] Fine (1974, pp. 261–4) shows the equivalence between the product rule and this assumption on joint distributions, given the spectrum rule of note 2. The claim in the text follows from that.

side of eqn 5b and then reading eqn 5b from right to left, they will be correct.

The preceding construction is neither pretty nor simple. But it shows something. It shows that the Hardy theorem, whether in the probabilistic versions explored in section 3 or in the apparently simpler version about possessed values, depends on more than locality. It depends, in addition, on a tacit requirement for how to deploy the framework of random variables in building a hidden variables model: namely, that we should employ the joint distribution structure of the random variables and demand that where applicable it match the quantum joints. If you think about it, however, this requirement is not really compelling, since we know in advance that the match can at best be partial. For all pairs of random variables have joint distribution but only some pairs of the quantum observable do. Thus any deployment of random variables necessarily involves excess structure.

The preceding construction shows something else too. It shows how to build a local hidden variables model for the Hardy example. All we need do is to make the suggested construction in the given state Ψ for the observables A, A', B, B', AB, AB', $A'B$, $A'B'$, $I-A$ and $I-B$, $A'(I-B)$ and $(I-A)B$. This will give a possessed value to each observable and probabilities that match those of quantum mechanics, as in eqn 2. We may find, for some 'hidden states', that $A'=1$ and $B'=1$, that $A=1$ but that $B=0$ and $(I-B)=1$, even though eqn 2b – $P^{\Psi}(B|A')=1$– holds. For eqn 2b is equivalent to $P^{\Psi}(A'\bar{B})=0=P^{\Psi}[A'(I-B)]$. The right side of course requires that the product $[A'(I-B)]=0$ but this is now compatible with $A'=1$ and $(I-B)=1$, since we no longer insist on eqn 5a and hence on the product rule. I said it was not pretty. But it is local, respects the probabilities of quantum mechanics and can accommodate all the measurement results. The 'funny' values – the ones that violate the product rule, or the like – can be regarded as values assigned to observables that do not commute with the ones being measured. From the quantum mechanical point of view these values are truly hidden, but they are locally assigned nevertheless.

The counterfactual reasoning that usually supports the interpretation of the Hardy theorem as a nonlocality proof goes wrong at the very start. Even before it gets entangled in nested counterfactuals it assumes that if a measurement turns up $A'=1$, then by virtue of $P^{\Psi}(B|A')=1$ we must have that $B=1$. We see above, however, that locality alone does not make this necessary; that is, not unless B is co-measured with A', in which case it will follow (just as it does in quantum mechanics). Reference to 'elements of reality' here – with its historically misleading echoes of EPR[4] – is equally off the track. For the quantum probability assignment $P^{\Psi}(B|A') = 1$ does not say that we can predict the value $B = 1$ from any measurement yielding $A' = 1$ that does not disturb the B system. It says that if we measure A' and B *together*, then where we find that $A' = 1$ we also find that $B = 1$. So even if we follow the prescription for elements of reality, all we can say is that where we measure A' and B together and find $A' = 1$ we can assign a $B = 1$ 'element of reality'. The Hardy argument, however, begins by supposing we have measured A' with B', not with B. In this situation there is no $B = 1$ 'element of reality' at all.

Beneath all these sophisticated arguments, I suggest, is a very simple conception for how we 'should' assign values when we try to respect locality and how we 'should' make that match up with the quantum probabilities. That conception is the random variables framework with the assumption of eqn 5a, or some equivalent. My point is that this conception goes well beyond the commitments of locality, which can be salvaged by assigning values differently. That means that the Hardy theorem, like other variants on Bell, is not a 'proof of nonlocality'. It is a proof that locality cannot be married to the assignment of determinate values in the recommended way. That is interesting and significant. It is not, however, a demonstration that quantum mechanics is nonlocal, much less (as some proclaim) that nature is.

References

Bell, J. (1966). 'On the Problem of Hidden Variables in Quantum Mechanics', *Review of Modern Physics* **38**, 447–75.

Beller, M. & A. Fine (1993). 'Bohr's Response to EPR', in J. Faye & H. Folse (eds), *Niels Bohr and Contemporary Philosophy* (Dordrecht: Kluwer), pp. 1–31.

Bohr, N. (1935). 'Can Quantum Mechanical Description of Physical Reality be Considered Complete?', *Physical Review* **48**, 696–702.

Einstein, A., B. Podolsky & N. Rosen (1935). 'Can Quantum Mechanical Description of Physical Reality be Considered Complete?', *Physical Review* **17**, 777–80.

[4] I am sure that the interest in 'elements of reality' derives from their association with Einstein (the 'E' of EPR). The 'reality criterion' that governs the introduction of these elements in EPR, however, is almost certainly due to Podolsky, who wrote the paper. In EPR that criterion plays a subtle and minor role, quite different from its use in the Hardy literature and other recent writings on Bell's theorem without inequalities. That use, rather, is a descendant of the apocryphal version of EPR made up by Bohr (1935). See Beller & Fine (1993) for details.

Fine, A. (1974). 'On The Completeness of Quantum Theory', *Synthese* **29**, 257–89. Reprinted with addendum in P. Suppes (ed.), *Logic and Probability in Quantum Mechanics*. (Dordrecht: D. Reidel), pp. 249–81.

(1982a). 'Hidden Variables, Joint Probability and the Bell Inequalities', *Physical Review Letters* **48**, 291–5.

(1982b). 'Joint Distributions, Quantum Correlations and Commuting Observables', *Journal of Mathematical Physics* **23**, 1306–10.

Goldstein, S. (1994). 'Nonlocality without Inequalities for Almost All Entangled States for Two Particles', *Physical Review Letters* **72**, 1951.

Greenberger, D. M., M. A. Horne & A. Zeilinger (1990). 'Going Beyond Bell's Theorem', in M. Kafatos (ed.), *Bell's Theorem, Quantum Theory, and Conceptions of the Universe*. (Dordrecht: Kluwer), pp. 69–72.

Hardy, L. (1993). 'Nonlocality for Two Particles without Inequalities for Almost All Entangled States', *Physical Review Letters* **71**, 1665–8.

Heywood, P. & M. L. G. Redhead (1983). 'Nonlocality and the Kochen–Specker Paradox', *Foundations of Physics* **13**, 481–99.

Kochen, S. & E. Specker (1967). 'The Problem of Hidden Variables in Quantum Mechanics', *Journal of Mathematics and Mechanics* **17**, 59–87.

Peres, A. (1990). 'Incompatible Results of Quantum Measurements', *Physics Letters A* **151**, 107–8.

Redhead, M. (1987). *Incompleteness, Nonlocality and Realism* (Oxford: Clarendon Press).

2

Beables in algebraic quantum theory

ROB CLIFTON

1. 'Observables' versus *be*ables

A good deal of Redhead's *Incompleteness, Nonlocality, and Realism* (1989) focuses on whether the no-go theorems of Kochen and Specker, and of Bell, make it impossible to maintain a 'simple realism of possessed values' (1989, p. 136) with regard to quantum-mechanical observables. No doubt Redhead's views on the matter remain tentative and exploratory, but in his more recent book *From Physics to Metaphysics* (1995a, ch. 3) he appears to favour van Fraassen's (1973) idea of securing simultaneously determinate values for all quantum-mechanical observables of a system by 'ontologically contextualizing' its physical magnitudes. The idea is to let any given degenerate self-adjoint operator on a system's Hilbert space represent more than one magnitude of the system. Each magnitude is distinguished from the others by the functional relations its values have to different complete commuting sets of self-adjoint operators of which the given self-adjoint operator is a member. Thus, to pick out a physical magnitude it is not enough to know that its statistics are represented by tracing the density operator of the system with some particular self-adjoint operator; for the degenerate operators, one must also pick a context of definition for the magnitude being measured, specified by some complete commuting set.

Formally, this is enough to prevent Kochen–Specker contradictions. But for Redhead the real payoff is that ontological contextuality yields a novel holistic interpretation of quantum nonlocality (1995a, pp. 86–7). Take the example of two correlated spin-1 particles in the entangled singlet state

$$-3^{-\frac{1}{2}}[|S_{1x}=0\rangle|S_{2x}=0\rangle - |S_{1y}=0\rangle|S_{2y}=0\rangle + |S_{1z}=0\rangle|S_{2z}=0\rangle] \qquad (1)$$

I would like to thank John L. Bell, Alex Wilce, and especially Hans Halvorson for occasional guidance on matters mathematical. I also thank Klaas Landsman, David Malament, and Laura Ruetsche for their useful feedback on the first draft of this essay.

for which Heywood and Redhead (1983) were able to supply the first purely algebraic proof of Bell's nonlocality theorem.[1] Since the self-adjoint operator S_{1n}^2 pertaining to any (squared) spin component n of particle 1 is represented by the degenerate operator $S_{1n}^2 \otimes I_2$ on $1 + 2$'s Hilbert space (where I_2 is the identity operator on 2's space), ontological contextualism blocks the conclusion – which would otherwise be forced by the Heywood–Redhead argument – that the outcome of a measurement of S_{1n}^2 must causally depend upon measurements performed on particle 2. This conclusion is blocked because the very definition of the spin magnitude being measured rests on which complete commuting set of self-adjoint operators of the composite $1 + 2$ system the values of $S_{1n}^2 \otimes I_2$ are referred to. And if we cannot specify a subsystem's properties independently of properties relating to the whole combined system, then the question of whether properties intrinsic to a subsystem causally depend upon measurements undertaken on spacelike-separated systems cannot even be raised. Redhead calls this consequence of ontological contextualism 'ontological nonlocality' to contrast it with 'environmental nonlocality' that would involve an explicit spacelike causal dependence of local properties on distant measurements in apparent conflict with relativity theory.

Despite the lure of this route to peaceful coexistence between relativity and quantum nonlocality, it is hard to be totally at ease with an ontology that entertains the existence of large numbers of distinct physical magnitudes which are, nevertheless, statistically indistinguishable (in principle). And since it is far from obvious how failing to classify quantum nonlocality as a causal connection improves the chances of securing a Lorentz invariant realist interpretation of quantum theory, it is surely worth seeking an alternative, *simpler* realism of possessed values that takes the functional relations between self-adjoint operators just as seriously.

In fact, one does not have to look very far. The key lies in rejecting an assumption that is necessary to prove the Kochen–Specker theorem which Redhead dubs the 'Reality Principle' in *Incompleteness, Nonlocality and Realism* (1987): 'If there is an operationally defined number associated with the self-adjoint operator $\hat{Q}$ (i.e. distributed probabilistically according to the statistical algorithm of QM for $\hat{Q}$), then there exists an element of reality . . . associated with that number and measured by it' (1987, p. 133–4). Redhead considers (and rightly rejects) only one way to deny the Reality Principle. Faced with incompatible ways to measure a degenerate self-

[1] See also Redhead (1989), ch. 6.

adjoint operator $\hat{Q}$ – depending on which complete commuting set it is measured along with – one could say that only one way reveals $\hat{Q}$'s true value and the others 'produce numbers which just "hang in the air" and do not measure anything of ontological significance' (1987, p. 136). But this is not the most natural way to deny the Reality Principle. The most natural way is to regard the measurement of certain self-adjoint operators as yielding results without ontological significance *however* they are 'measured' – the paradigm example being the measurement of 'spin' in Bohm's theory.[2] This is not at all to renounce the realist demand for an explanation of measurement results, but merely to abandon the particular form of explanation demanded by the Reality Principle, which dictates that each self-adjoint operator needs to be thought of as having its measurement results determined by a pre-existing element of reality *unique to that operator*. (In Bohm's theory, by contrast, all measurement results, regardless of what self-adjoint operator has been 'measured', are explained in terms of the pre-existing *position* of the particle, its initial wavefunction, and the theory's dynamics.) Jettisoning this part of the Reality Principle clears the way for an interpretative programme in quantum mechanics which has received concrete expression recently in various 'modal' interpretations of quantum mechanics,[3] and has figured prominently in the writings of Bell, collected together in his book *Speakable and Unspeakable in Quantum Mechanics* (1987, particularly chs. 5, 7, 19).

For Bell, a self-adjoint operator is *prima facie* just a mathematical device which, when traced with the system's density operator, generates the empirically correct statistics in an experiment on the system which orthodox quantum mechanics would loosely call a 'measurement' of the 'observable' represented by the operator (1987, p. 52). Out of the 'observables' of the orthodox interpretation Bell seeks to isolate some subset, the '*be*ables' of a system, that can be ascribed determinate values and about which orthodoxy's loose talk is perfectly precise:

> Many people must have thought along the following lines. Could one not just promote *some* of the observables of the present quantum theory to the status of beables? The beables would then be represented by linear operators in the state space. The values which they are allowed to *be* would be the eigenvalues of those operators. For the general state the probability of a beable *being* a particular value would be calculated just as was formerly calculated the probability of *observing* that value. (1987, p. 41)

[2] See Pagonis & Clifton (1995) and Dürr et al. (1996) for further discussion.

[3] See Clifton (1996), Bub (1997), and references therein.

Bell's thinking is the exact opposite of Redhead's in *From Physics to Metaphysics*. While Redhead entertains the bold conjecture that there are far more beables than self-adjoint operators, Bell is content with there being far less. Elsewhere, Bell explains how one can get away with this:

> Not all 'observables' can be given beable status, for they do not all have simultaneous eigenvalues, i.e. do not all commute. It is important to realize therefore that most of these 'observables' are entirely redundant. What is essential is to be able to define the positions of things, including the positions of instrument pointers or (the modern equivalent) of ink on computer output. (1987, p. 175)

> 'Observables' must be *made*, somehow, out of beables. The theory of local beables should contain, and give precise physical meaning to, the algebra of local observables. (1987, p. 52)

In the last passage above we see a further interesting divergence from Redhead's approach. While his ontological nonlocality countenances locally measurable but nonlocally defined beables (in order to avoid the spectre of spacelike causation), Bell restricts his considerations to beables that are both locally measurable *and* locally defined.[4] In fact, in two separate places (1987, chs. 5, 7) Bell expresses a strong interest in modelling a theory of local beables after Haag's (1992) algebras of local observables in relativistic quantum field theory. For example, the opening lines of Bell's 'The Theory of Local Beables' read:

> This is a pretentious name for a theory which hardly exists otherwise, but which ought to exist. The name is deliberately modelled on 'the algebra of local observables'. The terminology, *be*-able as against *observ*-able, is not designed to frighten with metaphysic those dedicated to realphysic. It is chosen rather to help in making explicit some notions already implicit in, and basic to, ordinary quantum theory. For, in the words of Bohr, 'it is decisive to recognize that, however far the phenomena transcend the scope of classical physical explanation, the account of all evidence must be expressed in classical terms'. It is the ambition of the theory of local beables to bring these 'classical terms' into the equations, and not relegate them entirely to the surrounding talk. (1987, p. 52)

In view of the C^*-algebraic formulation of Haag's theory of local observables, the ambition Bell expresses in the final sentence above (combined with Redhead's ambition to preserve functional relations between beables), suggests the following problem: provide an algebraic characterization of those *subsets* of (bounded) observables which are viable candidates for representing the *beables* of a quantum-mechanical system – be it a spin-1 particle in

[4] See also pp. 42 and 53 of Bell's (1987).

the singlet state of eqn 1, or a bounded open region of spacetime. My aim in the present essay is to pursue this problem while introducing those unfamiliar with algebraic quantum theory to the powerful resources it has to solve it.

I begin in section 2 by considering what subsets of the self-adjoint part of the C^*-algebra $\mathcal{U}$ of a quantum system should be candidates for beable status. No doubt this is partly a matter of taste. But a natural requirement to impose is that sets of beables be closed under the taking of any continuous self-adjoint function of their members. As we shall see, such sets have their own characteristic algebraic structure, and I call them 'Segalgebras' because they conform to the general postulates for algebraic systems of observables laid down and studied by Segal (1947). It turns out that a subset of self-adjoint elements in $\mathcal{U}$ forms a Segalgebra exactly when it is the self-adjoint part of some C^*- *sub*algebra of $\mathcal{U}$. This is satisfying insofar as there is no reason to expect that a set of beables should have an algebraic structure any different from the full set of observables of the system out of which the beables are distinguished. And the fact that Segalgebras are none other than the self-adjoint parts of C^*-algebras will allow us to carry over well-known facts about C^*-algebras to their Segalgebras.[5]

In section 3 I discuss what is going to count as an acceptable way of assigning values to beables in a Segalgebra. Again this is partly a matter of taste, and has been hotly debated ever since von Neumann proved his infamous no-hidden-variables theorem. Nevertheless, I shall argue that while von Neumann's conception of values as given by linear functionals on the Segalgebra of 'observables' of a system is utterly inappropriate if their assigned 'values' are only understood dispositionally or counterfactually (so I agree completely with Bell 1987, ch. 1), there is no reason not to require the *categorically possessed* values of beables to be given by linear functionals, regardless of whether they commute. This will not lead in the direction of an algebraic analogue of von Neumann's no-go theorem, such as is proved by Misra (1967), simply because I shall not be supposing that all 'observables' have beable status!

[5] Klaas Landsman has called my attention to the fact that it is possible to define Segalgebras independently of C^*-algebras (not assuming, as I do, that their elements are drawn from the self-adjoint part of a C^*-algebra), and then prove that every Segalgebra, initially defined in the abstract, is in fact isomorphic to the self-adjoint part of some C^*-algebra. (Landsman (1998, ch. 1) calls the abstract counterpart to my Segalgebras 'Jordan–Lie–Banach' algebras, and discusses their structure and the role they play within algebraic foundations of quantum theory.)

In line with the approach to value definiteness taken by modal interpretations, I will also not be requiring that the beables of a quantum system be the same from one quantum state of the system to another. In the second passage from Bell (1987) quoted above he implies that the self-adjoint operators corresponding to beables must all commute. It will turn out that this only follows if one requires the beables of a system to be the same for all the quantum states it could possibly occupy over time (as occurs in Bohm's theory). However, even if that requirement is dropped, we shall see that a Segalgebra of beables still has to be 'almost commutative'.

In section 4 I shall introduce the new notion of a quasicommutative Segalgebra to make this idea precise. In section 5 I go on to show that quasicommutative Segalgebras are both necessary and sufficient for representing the measurement statistics prescribed by a quantum state as an average over the actual values of the beables in the algebra[6] – in line with the first passage from Bell (1987) quoted above. I also discuss two concrete examples of non-commuting Segalgebras of beables employed by the orthodox (Dirac–von Neumann) interpretation of non-relativistic quantum mechanics and modal interpretations thereof. Section 6 then discusses the question of how 'big' (and non-commutative) a Segalgebra of beables can be consistent with satisfying the statistics of some quantum state. There is a simple characterization of the Segalgebras of beables on a finite-dimensional Hilbert space that are maximal in this sense. The analysis of the infinite-dimensional case requires more sophisticated techniques, and will appear in a later essay.

Finally, in section 7 I discuss two ways one can argue for entertaining only commutative Segalgebras of beables. The first way, as I have already mentioned, is to require that the beables of a system be the same for all its quantum states (or at least for a 'full set' thereof; we will see that the former demand would be too strong to impose, on physical grounds). The second way arises in the context of algebraic relativistic quantum field theory, where it turns out that Segalgebras of local beables must be *fully* commutative if they are to satisfy the measurement statistics dictated by any state of a quantum field on Minkowski spacetime that has bounded energy.

[6] This result generalizes results obtained by Bub & Clifton (1996) and Zimba & Clifton (1998) in two directions. First and most importantly, I shall not need to restrict myself to sets of beables with discrete spectra. (In particular, my results are completely consistent with Bohm's theory, which grants beable status to the configuration operator of a system, though some complications arise because that operator is unbounded – see section 5's discussion.) And, secondly, I shall not need to assume anything about the projection operators with beable status (such as: that they form an ortholattice); indeed, a C^*-algebra need have *no* non-trivial projections.

2. Segalgebras of beables

A $C^\star$-algebra is a normed algebra $\mathcal{U}$ over the complex numbers which is complete in the metric topology induced by the norm $|\cdot|$ and equipped with an involution * that, together with the norm, satisfies the *$C^\star$-norm property*,[7]

$$|A^\star A| = |A|^2, \text{ for all } A \in \mathcal{U}. \tag{2}$$

We will hardly ever need to suppose that our $C^\star$-algebras are concrete algebras of bounded operators acting on some Hilbert space, but for convenience I will still refer to the elements of $\mathcal{U}$ as operators. Of course, for a quantum system represented by some $\mathcal{U}$, only the operators in $\mathcal{U}$'s self-adjoint part – consisting of all $A \in \mathcal{U}$ such that $A = A^\star$ – can represent the bounded *observables* of the system. For simplicity, I set aside the possibility of superselection rules and take the term 'observable' to be synonomous with 'self-adjoint operator'. Our task, then, is to lay down some natural guidelines for granting observables in $\mathcal{U}$ beable status.

It seems reasonable to require that a quantum system be such that its beables combine algebraically to yield other beables, the idea being that if a set of observables have definite values, any (or, at least, any reasonably well-behaved) self-adjoint function of them ought to have a definite value as well. Thus a set of beables should at least form a real vector space. And, starting with any single beable, one should be able to form polynomials over the reals in that beable which are also beables of the system. If we are going to allow these polynomials to have a constant term, we had better also require that the identity operator I be a beable.[8] Finally, it seems reasonable to require that sets of beables not just be closed under polynomial functions of their members, but all continuous functions thereof. Thus if an observable A is a beable, then $\sin A$, e^A, and (if the spectrum of A consists only of non-negative values) $\sqrt{A}$ should all have beable status too. There is only one reasonable way to define a non-polynomial continuous function of a bounded observable A, viz. as the norm limit of a sequence of polynomials in A by analogy with the Weierstrass approximation theorem from ordinary analysis.[9] Thus sets of beables will need to be closed in norm.[10]

[7] Note that this property, together with the triangle inequality and product inequality $|AB| \leq |A||B|$, entails that $|A| = |A^\star|$ for any $A \in \mathcal{U}$.

[8] Not all $C^\star$-algebras have an identity, but one can always be 'adjoined' to any $C^\star$-algebra – see Bratteli & Robinson (1987), prop. 2.1.5.

[9] Geroch (1985, ch. 52) contains a complete discussion.

[10] This is consistent with our beables remaining self-adjoint, since the limit of a sequence of self-adjoint operators must itself be self-adjoint due to the continuity of the adjoint operation (which follows from $|A^\star - B^\star| = |(A - B)^\star| = |A - B|$).

Of course we cannot require sets of beables to be closed under products, since the product of two observables is an observable (i.e. self-adjoint) only if they commute. However, we can always introduce a new symmetric product on $\mathcal{U}$ by

$$A \circ B \equiv 1/4[(A+B)^2 - (A-B)^2] = 1/2[A, B]_+, \tag{3}$$

which is manifestly such that if both A and B are self-adjoint $A \circ B$ will be too. Since the symmetric product of two self-adjoint operators is expressible, as above, in terms of real linear combinations and squares of self-adjoint operators, it follows from our requirements on sets of beables that they *are* closed under the symmetric product.

It is easy to see that the symmetric product on $\mathcal{U}$ is homogeneous (i.e. $r(A \circ B) = (rA) \circ B = A \circ (rB)$ for any real number $r \in \mathcal{R}$) and distributive over addition. Moreover, the symmetric product will be associative on any triple of elements $A, B, C \in \mathcal{U}$ if they mutually commute with respect to their $C^\star$ products (for in that case $\circ$ just reduces to the $C^\star$ product, which of course is associative by definition). However, if a triple of elements do not mutually commute, the symmetric product cannot be assumed associative. A simple example is provided by the $C^\star$-algebra $\mathcal{U}(H_2)$ of all Hermitian operators on complex two-dimensional Hilbert space. If we consider the Pauli spin operators σ_x and σ_y, then since they anti-commute we have $\sigma_x \circ \sigma_y = 0$, thus $\sigma_x \circ (\sigma_x \circ \sigma_y) = 0$; yet $(\sigma_x \circ \sigma_x) \circ \sigma_y = \sigma_x^2 \circ \sigma_y = I \circ \sigma_y = \sigma_y$.

Since raising elements in $\mathcal{U}$ to any desired $C^\star$ power can be re-expressed in terms of the symmetric product as

$$A^n = \underbrace{A \circ A \circ \cdots \circ A}_{n \text{ times}}, \tag{4}$$

we can dispense with reference to the $C^\star$ product in our requirements on beable sets. Thus what we have required, so far, is that any set of beables be a real closed linear subspace of observables taken from $\mathcal{U}$ which forms a (not necessarily associative!) algebra with respect to the symmetric product. This is an instance of the sort of algebraic structure studied by Segal (1947), and a simple concrete example is given by

$$\{a\sigma_x + b\sigma_y + cI \in \mathcal{U}(H_2) | a, b, c \in \mathcal{R}\}. \tag{5}$$

There is one last requirement I need to impose on beable sets. We can also introduce on $\mathcal{U}$ an *anti*symmetric product given by

$$A \bullet B \equiv i/2[A, B]_-, \tag{6}$$

which also has the property that if both A and B are self-adjoint so is $A \bullet B$. This product is again homogeneous and distributive but not necessarily associative (e.g. $\sigma_x \bullet (\sigma_x \bullet \sigma_y) = -\sigma_y$, while $(\sigma_x \bullet \sigma_x) \bullet \sigma_y = 0$). If we are serious about wanting sets of beables to contain *all* continuous self-adjoint functions of their members, then they ought to be closed under the antisymmetric product too (its continuity is proven using the triangle and product inequalities). With closure under both the symmetric and antisymmetric products, we also obtain closure under self-adjoint polynomials in two beables, such as

$$cAB + c^*BA = 2\Re(c)A \circ B + 2\Im(c)A \bullet B. \tag{7}$$

It must be admitted, though, that closure under $\bullet$ is a strong assumption; for example, the set in eqn 5 is now ruled out, since it fails to contain $\sigma_x \bullet \sigma_y = -\sigma_z$. It might be of interest to investigate what portion of my conclusions can be recovered without assuming sets of beables are closed under $\bullet$, but I shall not do so here.

To summarize, our candidate beable sets are to be real norm-closed linear subspaces of observables in $\mathcal{U}$ that contain the identity and are closed under the (generally non-associative) symmetric and antisymmetric products. Such structures I call *Segalgebras* to distinguish them within the class of Segal's own algebras, which need not admit an antisymmetric product.[11] Virtually everything about Segalgebras follows from the fact that they are simply the self-adjoint parts of $C^\star$-subalgebras of the $C^\star$-algebras from which their elements are drawn.

To see this, recall that a subalgebra of a $C^\star$-algebra $\mathcal{U}$ is a subset of the algebra (possibly not containing the identity) that is closed under the relevant operations, i.e. a complex norm-closed subspace of $\mathcal{U}$ closed under the taking of $C^\star$ products and adjoints. For any set of observables T in $\mathcal{U}$, define

$$T + iT = \{A \in \mathcal{U} | A = X + iY, \text{ with } X, Y \in T\}. \tag{8}$$

Then we have:

Theorem 1 A subset T of the observables in a $C^\star$-algebra $\mathcal{U}$ is a real closed linear subspace of $\mathcal{U}$ closed under the symmetric and antisymmetric products if and only if $T + iT$ is a $C^\star$-subalgebra of $\mathcal{U}$.

Proof. 'If'. Assuming $T + iT$ is a subalgebra of $\mathcal{U}$, it is automatic that T is a real linear subspace. Moreover, since any Cauchy sequence $\{A_n\} \subseteq T$ must at least converge to an element $A \in T + iT$, and the limit of a sequence of self-adjoint elements must itself be self-adjoint, A must lie in the self-adjoint

[11] Neither does Segal's (1947) symmetric product (which he calls 'formal product') have to be homogeneous or distributive!

part of $T + iT$, which is obviously T. Thus T is norm-closed. Now recall that if D is an element in a $C^\star$-algebra, it has unique real and imaginary parts given by

$$\Re(D) = \frac{1}{2}(D + D^\star), \ \Im(D) = \frac{1}{2}(-iD + iD^\star). \tag{9}$$

To prove T is closed under symmetric and antisymmetric products, suppose $A, B \in T$. Then $A, B \in T + iT$, and since $T + iT$ is a subalgebra of $\mathcal{U}$, $AB \in T + iT$ has unique real and imaginary parts. Using eqns 9, those parts are just $A \circ B$ and $-A \bullet B$ and the conclusion follows.

'Only if'. Given that T is a real subspace of $\mathcal{U}$, it is routine to check that $T + iT$ is a complex subspace closed under *. Next, suppose $\{A_n\} \subseteq T + iT$ is a Cauchy sequence, i.e. $|A_n - A_m| \to 0$. From eqns 9, the triangle inequality, and the fact that $|D| = |D^\star|$ for any $D \in \mathcal{U}$, we see that $|\Re(D)|, |\Im(D)| \leq |D|$. Therefore,

$$|\Re(A_n) - \Re(A_m)| = |\Re(A_n - A_m)| \leq |A_n - A_m| \to 0, \tag{10}$$

and, similarly, $|\Im(A_n) - \Im(A_m)| \to 0$. So both $\{\Re(A_n)\}$ and $\{\Im(A_n)\}$ must be Cauchy sequences in T. Letting their respective limits be $A_1, A_2 \in T$, further use of the triangle inequality establishes that $A_1 + iA_2$ is the limit in $\mathcal{U}$ of $\{A_n\}$. Hence $T + iT$ is norm-closed. Finally, for closure of $T + iT$ under $C^\star$-products, let $A, B \in T + iT$, so

$$A = X + iY \text{ and } B = X' + iY' \text{ with } X, Y, X', Y' \in T. \tag{11}$$

A simple calculation yields

$$\Re(AB) = X \circ X' + X \bullet Y' + Y \bullet X' - Y \circ Y', \tag{12}$$

$$\Im(AB) = -X \bullet X' + X \circ Y' + Y \circ X' + Y \bullet Y'. \tag{13}$$

Therefore, since T is a real linear subspace closed under both the symmetric and antisymmetric products, both the real and imaginary parts of AB must lie in T, and therefore $AB \in T + iT$. *QED.*

Thm 1 tells us that a subset of $\mathcal{U}$ is a Segalgebra exactly when it is the self-adjoint part of some $C^\star$-subalgebra of $\mathcal{U}$ that contains the identity. As a first example of how this makes the 'theory' of Segalgebras parasitic upon facts about the $C^\star$-algebras they generate, consider the maps that preserve these structures. Recall that a mapping of $C^\star$-algebras $\psi : \mathcal{U} \to \mathcal{U}'$ is called a **-homomorphism* if it preserves the identity, linear combinations, products and adjoints. It is a theorem that *-homomorphisms are continuous,[12] so in

[12] Bratteli & Robinson (1987), prop. 2.3.1.

fact they preserve *all* the relevant structure of a $C^\star$-algebra. Analogously, call a mapping of Segalgebras $\phi : \mathcal{S} \to \mathcal{S}'$ a *homomorphism* if it preserves the identity, linear combinations, and symmetric and antisymmetric products. There is an obvious bijective correspondence between *-homomorphisms and homomorphisms. If $\psi : \mathcal{U} \to \mathcal{U}'$ is a *-homomorphism, the restriction of ψ to $\mathcal{U}$'s Segalgebra (i.e., to $\mathcal{U}$'s self-adjoint part) is a homomorphism into $\mathcal{U}'$'s. Conversely, if $\phi : \mathcal{S} \to \mathcal{S}'$ is a homomorphism, the (unique) linear extension of ϕ to $\mathcal{S} + i\mathcal{S}$ given by $\psi(A) = \phi(\Re(A)) + i\phi(\Im(A))$ is a *-homomorphism into $\mathcal{S}' + i\mathcal{S}'$. (To check that ψ preserves $C^\star$ products, use eqns 11–13.) Due to this bijective correspondence, we learn 'for free' that homomorphisms of Segalgebras must also be continuous.

3. Statistical states and value states

Having decided that our sets of beables will have the algebraic structure of Segalgebras, the next step is to decide how to assign values to beables. For this, we first need to recall the algebraic definition of a quantum state.

An operator A in a $C^\star$-algebra $\mathcal{U}$ is called *positive* if it is self-adjoint and has a non-negative spectrum. It is useful to have on hand two alternative equivalent definitions: A is positive if it is the square of a self-adjoint operator in $\mathcal{U}$, or if there is a $B \in \mathcal{U}$ such that $A = B^\star B$.[13] A *state* on a $C^\star$-algebra $\mathcal{U}$ is a (complex-valued) linear functional on $\mathcal{U}$ that maps positive operators to non-negative numbers and the identity to 1. It is a theorem that states, so defined, are continuous.[14]

We can define a *state on a Segalgebra* $\mathcal{S}$ in pretty much the same way, as a (this time, real-valued) linear functional on $\mathcal{S}$ that maps squares to non-negative numbers and I to 1. Again, there is the obvious bijective correspondence between states on Segalgebras and on the $C^\star$-algebras they generate, so that Segalgebraic states are continuous too. Now, in general, a state on a Segalgebra is merely statistical and specifies only the expectation values of the observables the state acts upon. If we want to interpret those expectation values as averages over the actual values of beables, we need a specification of the allowed valuations.

Certainly a valuation on the beables in a Segalgebra should itself be a real-valued function on it. Moreover, valuations ought to assign non-negative numbers to observables with non-negative spectra! And valuations

[13] See Bratteli & Robinson (1987), props. 2.2.10 and 2.2.12 respectively.
[14] Bratteli & Robinson (1987), prop. 2.3.11.

should map I to 1, if only because if some 'deviant' set of valuations did not, then since all the states whose statistics we want to recover with our valuations map I to 1, we would have to assign that deviant set measure-zero anyway. But should a valuation of the beables in a Segalgebra $\mathcal{S}$ be a *linear* function on $\mathcal{S}$? That is, should a valuation just be a special kind of state, in the technical sense above, that gives the actual values of beables rather than just their expectations? I believe that the assumption of linearity is defensible, yet I completely agree with Bell's (1987, ch. 1) critique of von Neumann's no-hidden-variables theorem (!). Let me explain.

What Bell objects to in both von Neumann's and Kochen and Specker's no-go theorems is arbitrary assumptions about how the results of measurements undertaken with incompatible experimental arrangements would turn out. For von Neumann, it is the assumption that if an observable C is actually measured, where $C = A + B$ and $[A, B] \neq 0$, then had A instead been measured, or B, their results would have been such as to sum to the value actually obtained for C. For Kochen and Specker, who adopt von Neumann's linearity requirement only when $[A, B] = 0$, it is the assumption that the results of measuring C would be the same independent of whether C is measured along with A and B or in the context of measuring some other pair of compatible observables A' and B' such that $C = A' + B'$. What makes these assumptions arbitrary, for Bell, is that the results of measuring observables $A, B, C, \ldots$ might not reveal separate pre-existing values for them (*contra* Redhead's Reality Principle, discussed in section 1 above), but rather realize *mere dispositions of the system to produce those results* in the context of the specific experimental arrangements they are obtained in. In other words, the 'observables' at issue need not all be *be*ables in the hidden-variables interpretations the no-go theorems seek to rule out.

Some commentators[15] place undue emphasis on the fact that, in his criticism of von Neumann's linearity assumption, Bell points out that it is mathematically possible for the measurement results dictated by hidden-variables in individual cases to violate linearity for noncommuting observables even while linearity of their expectation values is preserved after averaging over those variables. But it is wrong to portray Bell's critique as turning on a mere mathematical possibility, and it misses the reason why, for Bell, the theorems of von Neumann and Kochen–Specker stand or fall together. For having established the mathematical point beyond doubt using a simple toy hidden-variables model, Bell goes on to remark: 'At

[15] For example, Mermin (1993), sect. III.

first sight the required additivity of expectation values seems very reasonable, and it is rather the non-additivity of allowed values (eigenvalues) which requires explanation' (1987, p. 4). Bell then backs up this remark by giving a positive *physical* explanation for the non-additivity, in terms of measurement results displaying dispositions of the system in different experimental contexts rather than pre-existing values for the 'observables' measured.

Now while arbitrary assumptions about the results of measuring 'observables' are certainly to be avoided, it seems to me that there is no good physical reason (short of reintroducing some form of ontological contextuality) to reject linearity as a requirement on the categorically possessed values of *beables*. Of course, linearity for the values of noncommuting beables with discrete spectra is not going to be easy to satisfy, since the sum of any two eigenvalues for A and B need not even *be* an eigenvalue for $C = A + B$.[16] But rather than assign an eigenvalue to beable C in such cases that floats freely of the values assigned to beables A and B (yet on averaging linearity of expectation values is miraculously restored), it would be better not to have promoted 'observables' A and B to beable status in the first place![17] *That* is what I take the lesson of Bell's critique of the no-hidden-variables theorems to be.

There is one last requirement to impose on our beable valuations. The value assigned to the square of any beable should equal the square of the value assigned to that beable. It appears that Bohm's (1952) theory contradicts this by predicting that a particle confined to a box in a stationary state with definite energy $(1/2m)(nh/L)^2$ will possess zero momentum, so that its energy could not possibly be proportional to the square of the value of its momentum.[18] However, Bohm's theory also predicts that if the momentum of the particle is measured, it will be found to have non-zero values $\pm nh/L$ with equal probability. It follows that momentum in the Bohm theory is *not* a beable in Bell's sense, since the probability of *finding* a certain value for momentum is not the same as the probability that the particle *has* that momentum.[19]

[16] The standard example is $A = \sigma_x$, $B = \sigma_y$.

[17] One could avoid this conclusion by only withholding beable status from C, thereby rejecting my assumption that sets of beables form (real) linear spaces. But it would be hard to find principled reasons for that rejection. For example, the kinetic and potential energies of a particle in a potential will fail to commute. But if both were assumed to have pre-existing possessed values it would be difficult to make sense of the particle's total energy not having a possessed value too.

[18] Bohm (1952), sect. 15.

[19] Recall Bell's remark in the last sentence of the second quotation in sect. 1.

So we are requiring our beable value states to be, in the well-known jargon, dispersion-free states. A state ω on a Segalgebra $\mathcal{S}$ is called *dispersion-free on an observable* $A \in \mathcal{S}$ if $\omega(A^2) = (\omega(A))^2$, and a state is called *dispersion-free on* $\mathcal{S}$ if it is dispersion-free on *all* observables of $\mathcal{S}$. It is time now to develop some of the consequences of our assumption that valuations on Segalgebras of beables are given by dispersion-free states.

Obviously every homomorphism of $\mathcal{S}$ into the Segalgebra of real numbers $\mathcal{R}$ is a dispersion-free state on $\mathcal{S}$. (In $\mathcal{R}$ the symmetric product is the usual product, the antisymmetric product of any two real numbers is 0, and the norm of an element is its absolute value.) Conversely, every dispersion-free state on $\mathcal{S}$ is a homomorphism into $\mathcal{R}$. In fact, somewhat more is true, as a consequence of the following theorem.[20]

Theorem 2 Let ω be a state on a Segalgebra $\mathcal{S}$ that is dispersion-free on some $A \in \mathcal{S}$. Then $\omega(A)$ lies in the spectrum of A and for any $B \in \mathcal{S}$, $\omega(A \circ B) = \omega(A)\omega(B)$ and $\omega(A \bullet B) = 0$.

Proof. Extend ω to a state τ on the C^*-algebra $\mathcal{S} + i\mathcal{S}$ by defining $\tau(C) = \omega(\Re(C)) + i\omega(\Im(C))$. The map which sends each pair of elements $C, D \in \mathcal{S} + i\mathcal{S}$ to $\tau(C^\star D)$ defines a positive semi-definite inner product $\langle C|D\rangle$ on $\mathcal{S} + i\mathcal{S}$. Therefore, we can derive (in the usual way) the Schwartz inequality

$$|\tau(C^\star D)| \leq \sqrt{\tau(C^\star C)}\sqrt{\tau(D^\star D)}, \text{ for any } C, D \in \mathcal{S} + i\mathcal{S}. \tag{14}$$

Now since τ is dispersion-free on A, $\tau([A - \tau(A)I]^2) = 0$. Replacing C by $A - \tau(A)I$ in the Schwartz inequality above (and remembering that A is self-adjoint), it follows that $\tau([A - \tau(A)I]D) = 0$ for every $D \in \mathcal{S} + i\mathcal{S}$. Thus $\tau(A) = \omega(A)$ must lie in the spectrum of A; for were $A - \tau(A)I$ invertible in $\mathcal{S} + i\mathcal{S}$, we could set D equal to the inverse of $A - \tau(A)I$ and derive the contradiction $\tau(I) = 0$ from $\tau([A - \tau(A)I]D) = 0$.[21]

Continuing, since $\tau([A - \tau(A)I]D) = 0$ for every $D \in \mathcal{S} + i\mathcal{S}$, it follows that $\tau(AD) = \tau(A)\tau(D)$ for every such D. Using the same argument, we can replace D by $A - \tau(A)I$ and C by $D^\star$ in eqn 14 to obtain $\tau(DA) = \tau(D)\tau(A)$ for every $D \in \mathcal{S} + i\mathcal{S}$. Therefore, since τ agrees with ω on $\mathcal{S}$, for any $B \in \mathcal{S}$ we have

[20] The following result is a modification of standard arguments for C^*-algebras given in Kadison (1975, pp. 105–6).

[21] Since invertibility in $\mathcal{S} + i\mathcal{S}$ is equivalent to invertibility in any C^*-algebra with $\mathcal{S} + i\mathcal{S}$ as a subalgebra (Bratteli & Robinson 1987, prop. 2.2.7), this argument does not (and had better not!) assume that $\mathcal{S}$ is the entire self-adjoint part of the C^*-algebra describing some system.

$$\begin{aligned}\omega(A \circ B) &= \tau(1/2[AB + BA]) && (15)\\ &= 1/2[\tau(AB) + \tau(BA)] && (16)\\ &= 1/2[\tau(A)\tau(B) + \tau(B)\tau(A)] && (17)\\ &= 1/2[\omega(A)\omega(B) + \omega(B)\omega(A)] && (18)\\ &= \omega(A)\omega(B) && (19)\end{aligned}$$

and

$$\begin{aligned}\omega(A \bullet B) &= \tau(i/2[AB - BA]) && (20)\\ &= i/2[\tau(AB) - \tau(BA)] && (21)\\ &= i/2[\tau(A)\tau(B) - \tau(B)\tau(A)] = 0. && (22)\end{aligned}$$

QED.

Thm 2 is a mixed blessing. On the one hand, we only required from the outset that value states assign non-negative values to beables with non-negative spectra, so it is reassuring to now see that the value assigned to *any* beable must in fact lie *within* its spectrum. On the other hand, the theorem shows that asking for a Segalgebra of noncommuting beables with decently behaved valuations is a tall order. For suppose $[A, B]_- \neq 0$, so that $A \bullet B \neq 0$. Then since the spectrum of $A \bullet B$ must include $\pm|A \bullet B|$,[22] $A \bullet B$ must have at least one non-zero spectral value; yet thm 2 tells us that antisymmetric products of beables are always mapped to 0 by dispersion-free states!

It is exactly this sort of observation that leads Misra (1967, pp. 856–7) to conclude that hidden-variables in algebraic quantum mechanics are impossible. Working in the more general context of Segal's algebras, Misra introduces the idea of a derivation on the algebra, which is a linear mapping **D** from the algebra onto itself that satisfies the Leibniz rule with respect to the symmetric product:

$$\mathbf{D}(A \circ B) = A \circ \mathbf{D}(B) + \mathbf{D}(A) \circ B. \tag{23}$$

Misra does not assume the Segal algebras he works with form the self-adjoint parts of $C^\star$-algebras – that they are *Segalgebras* in my sense – though he *does* need to assume that every observable in the algebra is the difference of two positive observables in the algebra.[23] With that assumption, Misra (1967, thm 4) shows dispersion-free states map derivations to zero: for any A in the algebra, any derivation **D**, and any dispersion-free

[22] See Geroch (1985, thm 60).

[23] This is true for $C^\star$-algebras, and therefore Segalgebras – see prop. 2.2.11 of Bratteli & Robinson (1987).

state $\langle\cdot\rangle$, $\langle \mathbf{D}(A)\rangle = 0$. Since the antisymmetric product of any observable in a Segalgebra $\mathcal{S}$ with some given observable defines a derivation on $\mathcal{S}$, Misra's result entails thm 2's result that dispersion-free states on Segalgebras map antisymmetric products to zero. But, as we shall see explicitly in the final section of this paper, Misra's conclusion that his result 'excludes hidden variables in the general algebraic setting of quantum mechanics' (p. 857) is based upon a failure to distinguish 'observables' from beables. In fact, we shall shortly see that noncommuting Segalgebras of beables have *not* been excluded.

4. Quasicommutative Segalgebras

Clearly a Segalgebra $\mathcal{S}$ is *commutative* (in the usual $C^\star$ sense) if its antisymmetric product is trivial, i.e. $\mathcal{S} \bullet \mathcal{S} = \{0\}$. In this section I introduce the idea of a quasicommutative Segalgebra which we will see in the next section captures the precise extent to which beables can fail to commute. To adequately motivate and characterize 'quasicommutativity' from a formal point of view, I will first need an alternative characterization of commutativity and then I will need to discuss quotient Segalgebras.

There is a famous representation theorem for commutative $C^\star$-algebras from which an analogous result for commutative Segalgebras can be extracted. Recall that an archetypal example of a $C^\star$-algebra is the algebra $\mathcal{C}(X)$ of all complex-valued continuous functions on some compact Hausdorff topological space X. In $\mathcal{C}(X)$, linear combinations and products of functions are defined in the obvious (pointwise) way, the adjoint of a function is its complex conjugate, and the norm of a function is the maximum absolute value it takes over X.[24] The point about commutative $C^\star$-algebras is that they all arise in this way. *Every* commutative $C^\star$-algebra (with identity) is *-isomorphic to some $\mathcal{C}(X)$, for some compact Hausdorff space X.[25] Since a *-isomorphism of $C^\star$-algebras induces an isomorphism between their Segalgebras, it follows that every commutative Segalgebra $\mathcal{S}$ is isomorphic to the Segalgebra of all *real*-valued continuous functions $\mathcal{C}_R(X)$ on some compact Hausdorff space X.[26]

[24] $\mathcal{C}(X)$ is norm-closed, for the uniform limit of a sequence of (necessarily bounded) continuous functions on a compact space X must itself be a continuous function on X – see Simmons (1963, pp. 80–5).

[25] Bratteli & Robinson (1987), prop. 2.1.11A.

[26] $\mathcal{C}_{\mathcal{R}}(X)$ will just be the set of self-adjoint elements, i.e. real-valued functions, in the function algebra $\mathcal{C}(X)$ to which $\mathcal{S} + i\mathcal{S}$ is *-isomorphic. (See also Segal (1947), thm. 1.)

Now call a set of states Ω on a Segalgebra $\mathcal{S}$ *full* if Ω 'separates the points of $\mathcal{S}$' in the sense that, for any two distinct elements of $\mathcal{S}$, there is a state in Ω mapping them to different expectation values – or, equivalently (by the linearity of states), if for any non-zero $A \in \mathcal{S}$ there is a state $\omega \in \Omega$ such that $\omega(A) \neq 0$. Then the alternative characterization of commutativity I need is the following.[27]

Theorem 3 A Segalgebra is commutative if and only if it has a full set of dispersion-free states.

Proof. 'Only if'. If $\mathcal{S}$ is commutative it is isomorphic to the set of real-valued functions $\mathcal{C}_R(X)$ on some compact Hausdorff space X. Suppose $A \neq 0$. Then $A(x)$, the isomorphic image of A in $\mathcal{C}_R(X)$, cannot be the zero function (since the isomorphism can only map the zero operator to that function). So there is at least one point $x_0 \in X$ such that $A(x_0) \neq 0$. It is easy to see that the map defined by $\langle B\rangle = B(x_0)$ for all $B \in \mathcal{S}$ is a dispersion-free state on $\mathcal{S}$ satisfying $\langle A\rangle \neq 0$.

'If'. Consider any pair $A, B \in \mathcal{S}$ and their antisymmetric product $A \bullet B$. If $A \bullet B \neq 0$, then by hypothesis there is a dispersion-free state $\langle\cdot\rangle$ on $\mathcal{S}$ such that $\langle A \bullet B\rangle \neq 0$. But this contradicts thm 2, therefore $A \bullet B = 0$. *QED*.

In light of this result, the natural way to define quasicommutativity of a Segalgebra is in terms of it admitting a 'nearly full' set of dispersion-free states that only separate the points of $\mathcal{S}$ modulo some ideal in the algebra. It will turn out that factoring out the ideal yields a quotient Segalgebra that is (fully) commutative – which is what one might expect if the original (unfactored) Segalgebra were already 'close to being commutative'.

I shall introduce the idea of ideals and quotients for Segalgebras by again recalling their $C^\star$-algebra counterparts first. A (two-sided) ideal in a $C^\star$-algebra $\mathcal{U}$ is a subspace $\mathcal{I}$ of $\mathcal{U}$ which is invariant under multiplication on the left or right by any $A \in \mathcal{U}$, i.e. $\mathcal{UI} \subseteq \mathcal{I}$ and $\mathcal{IU} \subseteq \mathcal{I}$. We shall be interested only in closed *-ideals, i.e. ideals closed in norm and under the taking of adjoints. (An example is the collection of all functions in $\mathcal{C}(X)$ which vanish at some given point in X.) Clearly such an ideal is a subalgebra of $\mathcal{U}$. The idea is to 'factor out' this subalgebra so that what remains is again a $C^\star$-algebra. So as not to be left with something completely trivial, we will also require that $\mathcal{I}$ be *proper*, i.e., that $\mathcal{I}$ be a proper subset of $\mathcal{S}$, which is equivalent to requiring that $\mathcal{I}$ not contain the identity.

[27] This result is just a variation on Segal's (1947), thm 3.

Each proper closed *-ideal $\mathcal{I}$ in $\mathcal{U}$ determines an equivalence relation $\cong_{\mathcal{I}}$ on $\mathcal{U}$ defined by

$$A \cong_{\mathcal{I}} B \text{ if and only if } A - B \in \mathcal{I}. \tag{24}$$

The equivalence classes of $\cong_{\mathcal{I}}$ form a $C^\star$-algebra called the *quotient $C^\star$-algebra $\mathcal{U}/\mathcal{I}$ by the ideal $\mathcal{I}$*. To see how, let $\hat{A}$ denote the equivalence class in which A lies, and similarly for $\hat{B}$, $\hat{C}$, etc. Now define the relevant operations in $\mathcal{U}/\mathcal{I}$ by

$$c\hat{A} = \widehat{cA},\ \ \hat{A} + \hat{B} = \widehat{A + B},\ \ \hat{A}\hat{B} = \widehat{AB},\ \ \hat{A}^\star = \widehat{A^\star}, \tag{25}$$

noting that $\hat{I}$ is the identity in $\mathcal{U}/\mathcal{I}$ and $\hat{0}$ the zero element. (The factoring has been implemented by 'collapsing' everything in $\mathcal{I}$ into $\hat{0} \in \mathcal{U}/\mathcal{I}$.) Because $\mathcal{I}$ is a *-ideal, the definitions in eqn 25 are well-defined independent of the representatives chosen for the equivalence classes that appear in them. (For example, $\hat{A}\hat{B} = \widehat{AB}$ is well-defined only if $A \cong_{\mathcal{I}} A'$ and $B \cong_{\mathcal{I}} B'$ imply $AB \cong_{\mathcal{I}} A'B'$. But assuming the former, $(A - A')(B + B') \in \mathcal{I}$ and $(A + A')(B - B') \in \mathcal{I}$, therefore

$$\frac{1}{2}[(A - A')(B + B') + (A + A')(B - B')] = AB - A'B' \in \mathcal{I}.) \tag{26}$$

If we define

$$|\hat{A}| = \inf_{B \cong_{\mathcal{I}} A} |B|, \tag{27}$$

an elementary argument in the theory of Banach algebras establishes that $\mathcal{U}/\mathcal{I}$ is a complete normed algebra, and a not so elementary argument establishes the $C^\star$-norm property.[28]

The analogue of all this for Segalgebras may now be obvious. A proper closed ideal in $\mathcal{S}$ is a closed subspace $\mathcal{I}$ of $\mathcal{S}$ not containing the identity which is invariant under symmetric and antisymmetric multiplication, i.e. $\mathcal{S} \circ \mathcal{I} \subseteq \mathcal{I}$ and $\mathcal{S} \bullet \mathcal{I} \subseteq \mathcal{I}$. (Clearly we do not need to distinguish left from right multiplication.) By an argument virtually identical to the 'Only if' part of thm 1's proof, $\mathcal{I}$ extends to a proper closed *-ideal $\mathcal{I} + i\mathcal{I}$ in $\mathcal{S} + i\mathcal{S}$. So we can take the *quotient Segalgebra $\mathcal{S}/\mathcal{I}$ by the ideal $\mathcal{I}$* to be the Segalgebraic (i.e. self-adjoint) part of $(\mathcal{S} + i\mathcal{S})/(\mathcal{I} + i\mathcal{I})$. It is easy to see that for $A \in \mathcal{S}$, $\hat{A} \in \mathcal{S}/\mathcal{I}$ and that $A, B \in \mathcal{S}$ lie in the same equivalence class of $\mathcal{S}/\mathcal{I}$, i.e. $\hat{A} = \hat{B}$, if and only if $A \cong_{\mathcal{I}} B$. Furthermore, using eqns 25 it is easy to verify that the map which sends $A \in \mathcal{S}$ to $\hat{A} \in \mathcal{S}/\mathcal{I}$ is a homomorphism, and this homomorphism is surjective since $\hat{A} \in \mathcal{S}/\mathcal{I}$ is the image of $\Re(A) \in \mathcal{S}$.

[28] See Simmons (1963, thm 69D) and Bratteli and Robinson (1987, prop. 2.2.19), respectively.

We have now assembled all the necessary machinery to fulfil my promise to introduce a reasonable notion of quasicommutativity.

Theorem 4 Let $\mathcal{S}$ be a Segalgebra with proper closed ideal $\mathcal{I}$. Then the following are equivalent:

1. $\mathcal{S}/\mathcal{I}$ is commutative.
2. For any $A \notin \mathcal{I}$ there is a dispersion-free state $\langle\cdot\rangle$ on $\mathcal{S}$ such that $\langle A\rangle \neq 0$.
3. $\mathcal{S} \bullet \mathcal{S} \subseteq \mathcal{I}$.

Proof. 1. $\Rightarrow$ 2. If $A \notin \mathcal{I}$, then $\hat{A} \neq \hat{0}$. Since $\mathcal{S}/\mathcal{I}$ is commutative (by hypothesis), it has a full set of dispersion-free states (thm 3). So there is a dispersion-free state $\{\cdot\}$ on $\mathcal{S}/\mathcal{I}$ such that $\{\hat{A}\} \neq 0$. Defining $\langle B\rangle = \{\hat{B}\}$ for all $B \in \mathcal{S}$, the map $\langle\cdot\rangle$ – being the composition two homomorphisms, the second into $\mathcal{R}$ – is a dispersion-free state on $\mathcal{S}$ satisfying $\langle A\rangle \neq 0$.

2. $\Rightarrow$ 3. Identical to the 'If' part of thm 3 with $\mathcal{I}$ now playing the role of $\{0\}$.

3. $\Rightarrow$ 1. Let $\hat{A}, \hat{B} \in \mathcal{S}/\mathcal{I}$. Since $\mathcal{S} \bullet \mathcal{S} \subseteq \mathcal{I}$ (by hypothesis) and $\hat{\mathcal{I}} = \hat{0}$, we have $\hat{A} \bullet \hat{B} = \widehat{A \bullet B} = \hat{0}$, whence $\mathcal{S}/\mathcal{I}$ is commutative. *QED*.

In the case $\mathcal{I} = \{0\}$, parts 1 and 3 of thm 4 simply assert that $\mathcal{S}$ itself is commutative, and part 2 that $\mathcal{S}$ has a full set of dispersion-free states. So thm 4 generalizes thm 3 by relaxing its requirement of full commutativity. Motivated by this, call a Segalgebra $\mathcal{I}$-*quasicommutative* whenever it satisfies the equivalent conditions of thm 4.[29] Notice from 3. of thm 4 that if $\mathcal{S}$ is $\mathcal{I}$-quasicommutative and $\mathcal{J}$ is another proper ideal in $\mathcal{S}$ containing $\mathcal{I}$, then $\mathcal{S}$ is $\mathcal{J}$-quasicommutative as well. In particular, if $\mathcal{S}$ is commutative, it is automatically $\mathcal{I}$-quasicommutative with respect to any proper ideal $\mathcal{I} \in \mathcal{S}$. Since the converse fails, quasicommutativity is genuinely weaker than commutativity.

5. Beable subalgebras

It is high time I spelled out the connection between quasicommutativity and the problem of *beables* in quantum mechanics. For a Segalgebra of beables

[29] This idea deliberately parallels the idea of a 'quasidistributive lattice' introduced in Bell & Clifton (1995, thm 1). I should also note that results similar to 1. $\Leftrightarrow$ 2. of thm 4 have been proven by Misra (1967, thm 1) and Plymen (1968, thm 4.2) in the context of C^{*}- and von Neumann algebras, respectively.

to satisfy the statistics prescribed by some quantum state, we must be able to interpret the state's expectation values as averages over the actual values of the beables in the algebra. In other words, the quantum state must be a mixture of dispersion-free states on the algebra.

Let x be a variable in a measure space X, μ a positive measure on X such that $\mu(X) = 1$, and ω_x $(x \in X)$ a collection of states on a Segalgebra $\mathcal{S}$. Then the mapping defined by

$$\omega(A) = \int_X \omega_x(A) d\mu(x), \quad \text{for any } A \in \mathcal{S}, \tag{28}$$

will also be a state on $\mathcal{S}$. A state ω is called *mixed* if it can be represented, in the above way, as a weighted average of two or more (distinct) states with respect to some positive normalized measure μ; if it cannot, then ω is called *pure*. A subalgebra $\mathcal{B}$ of $\mathcal{S}$ will be said to *have beable status for the state* ω if $\omega|_{\mathcal{B}}$ – the restriction of ω to $\mathcal{B}$ – is either a mixture of dispersion-free states on $\mathcal{B}$ or is itself dispersion-free. As a check on the adequacy of this definition, we obtain the following intuitively expected result.[30]

Theorem 5 Every commutative subalgebra of $\mathcal{S}$ has beable status for every state on $\mathcal{S}$.

Proof. Let $\mathcal{C}$ be a commutative subalgebra of $\mathcal{S}$ and ω any of $\mathcal{S}$'s states. Since $\mathcal{C}$ is commutative, it is isomorphic to the set of real-valued functions $\mathcal{C}_{\mathcal{R}}(X)$ on some compact Hausdorff space X. Defining $\phi(A(x)) = \omega(A)$ for every $A \in \mathcal{C}$, ϕ is a state on $\mathcal{C}_{\mathcal{R}}(X)$. By the Riesz–Markov representation theorem,[31] ϕ must take the form

$$\phi(A(x)) = \int_X A(x) d\mu_\phi(x), \quad \text{for any } A(x) \in \mathcal{C}_{\mathcal{R}}(X), \tag{29}$$

where μ_ϕ is some positive normalized (completely additive) measure on X. But since for any $x \in X$ the map $\langle A \rangle_x = A(x)$ defines a dispersion-free state on $\mathcal{C}$, eqn 29 exhibits $\omega|_{\mathcal{C}}$ as a mixture of dispersion-free states on $\mathcal{C}$ (if $\omega|_{\mathcal{C}}$ is not already one of those states itself, which would correspond to the case where the complement of some point in X has μ_ϕ-measure zero). *QED*.

It is important to recognize that this theorem does actually cover the case of Bohm's theory. For simplicity, consider the space of states of a single spinless particle in one-dimension, given by the Hilbert space $L_2(\mathcal{R})$ of all (measurable) square-integrable, complex-valued functions on $\mathcal{R}$

[30] Here, I follow Segal's (1947, p. 933) argument almost to the letter.
[31] See Rudin (1974, thm 2.14).

(identifying functions that agree almost everywhere). The position $\hat{x}$ of the particle, and all self-adjoint functions thereof, are the only *bona fide* beables in Bohm's theory. Of course, a Segalgebra cannot contain any unbounded observables, and, in particular, could not contain $\hat{x}$. However, Bohm's theory certainly grants beable status to all *bounded* self-adjoint operator-valued functions of $\hat{x}$, and their (dispersion-free) values can be used to define a value for $\hat{x}$.

To see this, let f be any (measurable) essentially bounded, complex-valued function on $\mathcal{R}$, and define the bounded self-adjoint operator $\hat{O}_f$ by

$$\hat{O}_f(\psi(x)) = f(x)\psi(x) \text{ for each } \psi(x) \in L_2(\mathcal{R}). \tag{30}$$

The set of all such operators $\{\hat{O}_f\}$ is obviously commutative, and it is well-known that they form a $C^\star$-subalgebra of the $C^\star$-algebra of all bounded operators on $L_2(\mathcal{R})$.[32] If f is some essentially bounded function of x, $\hat{O}_f$ is the corresponding operator-valued function of $\hat{x}$ (keeping in mind that two fs that agree almost everywhere define the same operator $\hat{O}_f$). In fact, the operators $\{\hat{O}_f\}$ capture all the bounded operators which are functions of $\hat{x}$, since any such function would have to commute with all the $\hat{O}_f$s, and it is well-known that they form a maximal commutative set of bounded operators on $L_2(\mathcal{R})$.[33] Thus the subset of $\{\hat{O}_f\}$ where the fs are real-valued functions (almost everywhere) is the commutative sub-Segalgebra $\mathcal{S}_{\hat{x}}$ of all bounded observables that are functions of $\hat{x}$. In particular, the spectral projections of $\hat{x}$ correspond to characteristic functions $\chi(S)$ whose characteristic sets $S \subseteq \mathcal{R}$ have non-zero (Lebesgue) measure. Now let $\langle\cdot\rangle$ be a dispersion-free state on $\mathcal{S}_{\hat{x}}$, C any closed interval with *non-zero* length, and consider all projections in $\mathcal{S}_{\hat{x}}$ of the form $\hat{O}_{\chi(C)}$, representing the Yes/No questions: 'Does the particle lie in closed interval C?'. If there is no C such that $\langle\hat{O}_{\chi(C)}\rangle = 1$, we should not expect to be able to define the value of $\hat{x}$ relative to $\langle\cdot\rangle$. (Indeed, if a value for $\hat{x}$ were definable in *every* dispersion-free state on $\mathcal{S}_{\hat{x}}$, then since every state in $L_2(\mathcal{R})$ is a mixture of dispersion-free states on $\mathcal{S}_{\hat{x}}$, one could hope to define an expectation value for $\hat{x}$ in *every* state – contradicting the fact that $\hat{x}$ only has a well-defined expectation for the dense set of states in $L_2(\mathcal{R})$ lying in $\hat{x}$'s domain of definition.) But if there is some interval C with respect to which the answer to the corresponding Yes/No question, given $\langle\cdot\rangle$, is *Yes* (and recall that there is an abundant supply of dispersion-free states on $\mathcal{S}_{\hat{x}}$ with this property, since they consti-

32 See Geroch (1985, ch. 49) and Kadison & Ringrose (1983, sect. 4.1).

33 Kadison & Ringrose (1983, p. 308).

tute a full set), then $\hat{x}$'s value will be the *unique* point in the spectrum of $\hat{x}$ (i.e. the real line) lying in the set $S = \bigcap\{C : \langle \hat{O}_{\chi(C)} \rangle = 1\}$.[34]

To show that it is not necessary that subalgebras with beable status for some state be commutative, I need the following result.

Theorem 6 If ω is a state on a Segalgebra $\mathcal{S}$, the *null space* of ω defined by

$$\mathcal{N}_\omega = \{A \in \mathcal{S} | \omega(A^2) = 0\} \tag{31}$$

is a subalgebra of $\mathcal{S}$ on which ω is dispersion-free. Moreover, if $\mathcal{B} \subseteq \mathcal{S}$ has beable status for ω, then $\mathcal{N}_{\omega|_\mathcal{B}} = \mathcal{N}_\omega \cap \mathcal{B}$ is a proper closed ideal in $\mathcal{B}$.

Proof. As in the proof of thm 2, extend ω to a state τ on the $C^\star$-algebra $\mathcal{S} + i\mathcal{S}$. Fix an arbitrary $A \in \mathcal{N}_\omega$, so $\omega(A^2) = 0$. Replacing C by A in the Schwartz inequality (eqn 14) yields $\tau(AD) = 0$ for all $D \in \mathcal{S} + i\mathcal{S}$. In particular, with D replaced by I we see that $\tau(A) = \omega(A) = 0$, so that ω is dispersion-free on the set $\mathcal{N}_\omega$. Moreover, for any $A, B \in \mathcal{N}_\omega$ we have (using the definition of the symmetric product),

$$\begin{aligned} 2^{-2}\omega((A \circ B)^2) &= \omega([AB + BA][AB + BA]) && (32)\\ &= \tau(ABAB) + \tau(AB^2A) + \tau(BA^2B) + \tau(BABA) && (33)\\ &= 0, && (34) \end{aligned}$$

with a similar calculation applicable when $\circ$ is replaced by $\bullet$. It follows that $\mathcal{N}_\omega$ is closed under both $\circ$ and $\bullet$.

To see that $\mathcal{N}_\omega$ is a real norm-closed subspace of $\mathcal{S}$: norm-closure is immediate from the continuity of ω (and the 'square' operation), while assuming $A, B \in \mathcal{N}_\omega$ (with $a, b \in \mathcal{R}$) implies

$$\begin{aligned} \omega([aA + bB]^2) &= \omega([aA + bB] \circ [aA + bB]) && (35)\\ &= a^2\omega(A^2) + b^2\omega(B^2) + 2ab\omega(A \circ B) && (36)\\ &= 0 && (37) \end{aligned}$$

using thm 2 and the fact that ω assigns dispersion-free value zero to both A and B.

Finally, let $\mathcal{B} \subseteq \mathcal{S}$ have beable status for ω. Then $\mathcal{N}_{\omega|_\mathcal{B}}$, being the intersection of $\mathcal{N}_\omega$ with $\mathcal{B}$, is a subalgebra of $\mathcal{B}$ (which is of course proper, since $\omega(I) = 1$). It remains to show that $\mathcal{N}_{\omega|_\mathcal{B}}$ is an ideal. Fix an arbitrary $A \in \mathcal{N}_{\omega|_\mathcal{B}}$ and $B \in \mathcal{B}$. Since ω is a mixture of dispersion-free states $\langle \cdot \rangle_x$ $(x \in X)$ on $\mathcal{B}$, A^2

[34] The proof that S is non-empty and contains at most one point is given in Halvorson & Clifton (forthcoming, prop. 3.2), along with further discussion of the issues raised by value assignments to unbounded observables.

is positive, and $\omega(A^2) = 0$, $\langle A^2 \rangle_x = \langle A \rangle_x^2 = \langle A \rangle_x = 0$ for almost all $x \in X$. Therefore, for almost all x, $\langle (A \circ B)^2 \rangle_x = \langle A \circ B \rangle_x^2 = (\langle A \rangle_x \langle B \rangle_x)^2 = 0$; and, of course, for *all* x we have $\langle (A \bullet B)^2 \rangle_x = \langle A \bullet B \rangle_x^2 = 0$ (recalling that dispersion-free states are homomorphisms into $\mathcal{R}$). It follows that $\omega((A \circ B)^2) = \omega((A \bullet B)^2) = 0$, and $\mathcal{N}_{\omega|_{\mathcal{B}}}$ is a proper closed ideal. *QED.*

If one wants to include some pair of noncommuting observables A and B in a subalgebra with beable status for some state ω, the 'trick' is simply to make sure $A \bullet B$ lies inside the null space $\mathcal{N}_\omega$ determined by that state. If so, then it will not matter that dispersion-free states necessarily map antisymmetric products of noncommuting observables to zero, because ω *also* assigns definite value zero to $A \bullet B$! The following theorem formalizes this thought and supplies the promised connection between beable status and quasicommutativity.

Theorem 7 Let $\mathcal{B}$ be a subalgebra of a Segalgebra $\mathcal{S}$ and ω a state on $\mathcal{S}$. Then $\mathcal{B}$ has beable status for ω if and only if $\mathcal{B}$ is $\mathcal{N}_{\omega|_{\mathcal{B}}}$-quasicommutative.

Proof. 'Only if'. By thm 6, $\mathcal{N}_{\omega|_{\mathcal{B}}}$ is a proper closed ideal in $\mathcal{B}$. $\mathcal{B}$'s $\mathcal{N}_{\omega|_{\mathcal{B}}}$-quasicommutativity is then easily inferred from 2. of thm 4. Thus if $A \in \mathcal{B}$ with $A \notin \mathcal{N}_{\omega|_{\mathcal{B}}}$, then $\omega|_{\mathcal{B}}(A^2) \neq 0$. But by hypothesis, there is a collection of dispersion-free states $\{\langle \cdot \rangle_x | x \in X\}$ on $\mathcal{B}$ of which $\omega|_{\mathcal{B}}$ is a mixture. Therefore, for at least one $x_0 \in X$, $\langle A^2 \rangle_{x_0} \neq 0$ and hence $\langle A \rangle_{x_0} \neq 0$ as required.

'If'. By 1. of thm 4, $\mathcal{B}/\mathcal{N}_{\omega|_{\mathcal{B}}}$ is commutative. Define the map

$$\phi : \mathcal{B}/\mathcal{N}_{\omega|_{\mathcal{B}}} \rightarrow \mathcal{R} \text{ by } \phi(\hat{A}) = \omega|_{\mathcal{B}}(A). \tag{38}$$

Since the 'hat' map is surjective, this defines ϕ on all of $\mathcal{B}/\mathcal{N}_{\omega|_{\mathcal{B}}}$, and it is easy to check that ϕ is thereby well-defined. Furthermore, since the hat map is a homomorphism, ϕ is a state on $\mathcal{B}/\mathcal{N}_{\omega|_{\mathcal{B}}}$. By the commutativity of $\mathcal{B}/\mathcal{N}_{\omega|_{\mathcal{B}}}$ and thm 5, ϕ is a mixture of dispersion-free states $\langle \cdot \rangle_x$ $(x \in X)$ on $\mathcal{B}/\mathcal{N}_{\omega|_{\mathcal{B}}}$. But for any $\langle \cdot \rangle_x$ on $\mathcal{B}/\mathcal{N}_{\omega|_{\mathcal{B}}}$, $\langle \hat{\cdot} \rangle_x$ is a dispersion-free state on $\mathcal{B}$. So since $\omega|_{\mathcal{B}}(\cdot) = \phi(\hat{\cdot})$ (eqn 38), $\omega|_{\mathcal{B}}$ is a mixture of the dispersion-free states $\langle \hat{\cdot} \rangle_x$ $(x \in X)$ on $\mathcal{B}$. *QED.*

The examples of non-commutative subalgebras of beables that I shall construct make use of the following result.

Theorem 8 For any state ω on a Segalgebra $\mathcal{S}$, the *definite set* of ω defined by

$$\mathcal{D}_\omega = \{A \in \mathcal{S} | \omega(A^2) = (\omega(A))^2\} \tag{39}$$

has beable status for ω.

Proof. To see that $\mathcal{D}_\omega$ is a subalgebra, it is easiest to use the equivalence:

$$A \in \mathcal{D}_\omega \text{ if and only if } A - \omega(A)I \in \mathcal{N}_\omega, \tag{40}$$

and invoke the fact (thm 6) that $\mathcal{N}_\omega$ is a subalgebra of $\mathcal{S}$. To illustrate, let $A, B \in \mathcal{D}_\omega$. Then both $A - \omega(A)I$ and $B - \omega(B)I$ lie in $\mathcal{N}_\omega$, and therefore so does

$$(A - \omega(A)I) \circ (B - \omega(B)I) + \omega(A)[B - \omega(B)I] + \omega(B)[A - \omega(A)I] \tag{41}$$

which simplifies to $A \circ B - \omega(A)\omega(B)I \in \mathcal{N}_\omega$. But since $\omega(A)\omega(B) = \omega(A \circ B)$ by thm 2, this means $A \circ B \in \mathcal{D}_\omega$ (using eqn 40 once more).

The beable status of $\mathcal{D}_\omega$ for ω follows trivially from 2. of thm 4 and thm 7. Thus if A lies in $\mathcal{D}_\omega$ but not in $\mathcal{N}_\omega$, then of course there is a dispersion-free state on $\mathcal{D}_\omega$ mapping A to a non-zero value – ω itself! QED.

In the case where ω is a *vector state*, i.e. representable by a state vector $|v\rangle$ in a Hilbert space representation of the Segalgebra $\mathcal{S}$ (so $\omega(A) = \langle v|A|v\rangle$ for all $A \in \mathcal{S}$), ω's definite set consists of all those (bounded) self-adjoint operators on the Hilbert space that share the eigenstate $|v\rangle$ – a highly noncommutative set if the space has more than two dimensions. This is precisely the orthodox (Dirac–von Neumann) 'eigenstate–eigenvalue link' approach to assigning definite values to observables.

Definite sets can be used to build subalgebras with beable status for a state ω that are not simply subalgebras of ω's own definite set. The next result specifies a general class of examples of this sort, containing thm 8 as a degenerate case (when $\omega = \omega_x$ for all $x \in X$).

Theorem 9 Let $\mathcal{S}$ be a Segalgebra, ω a state on $\mathcal{S}$ and ω_x ($x \in X$) any family of states satisfying $\bigcap_{x\in X} \mathcal{N}_{\omega_x} \subseteq \mathcal{N}_\omega$. Then $\mathcal{B}_{\{\omega_x\}} = \bigcap_{x\in X} \mathcal{D}_{\omega_x}$ has beable status for ω.

Proof. Since $\mathcal{B}_{\{\omega_x\}}$ is the intersection of a collection of subalgebras of $\mathcal{S}$ (thm 8), it is itself a subalgebra. To establish beable status for ω, all we need to show (by 3. of thm 4 and thm 7) is that $\mathcal{B}_{\{\omega_x\}} \bullet \mathcal{B}_{\{\omega_x\}} \subseteq \mathcal{N}_\omega$. So suppose $A, B \in \mathcal{B}_{\{\omega_x\}}$. Then both A and B lie in the definite sets of all the ω_xs. As we know from eqn 40, this is equivalent to both $A - \omega_x(A)I$ and $B - \omega_x(B)I$ lying in each subalgebra $\mathcal{N}_{\omega_x}$, for all $x \in X$, which means

$$(A - \omega_x(A)I) \bullet (B - \omega_x(B)I) = A \bullet B \in \mathcal{N}_{\omega_x} \tag{42}$$

for all $x \in X$. But by hypothesis, $\bigcap_{x\in X} \mathcal{N}_{\omega_x} \subseteq \mathcal{N}_\omega$, therefore $A \bullet B \in \mathcal{N}_\omega$, which is the conclusion we were seeking. *QED*.

To obtain a concrete example of thm 9, consider again the case where ω is a vector state, represented by a state vector $|v\rangle$ in a Hilbert space representation H of $\mathcal{S}$. Let R be any bounded self-adjoint operator on H with discrete spectrum and eigenprojections $\{R_i\}$ – counting only those for which $R_i|v\rangle \neq 0$ – and consider the (renormalized) orthogonal projections $\{|v_{R_i}\rangle\}$ of the state vector $|v\rangle$ onto the eigenspaces of R. Since any observable with eigenvalue 0 in all the states $\{|v_{R_i}\rangle\}$ must have eigenvalue 0 in the state $|v\rangle$ (the latter lying in the span of the former), the conditions of thm 9 are satisfied so that $\mathcal{B}_{\{|v_{R_i}\rangle\}}$ has beable status in the state $|v\rangle$. This is the modal (i.e. state-dependent) method adopted by Bub (1997)[35] for building a set of beables out of the state $|v\rangle$ of a system and a 'preferred observable' R.[36]

Henceforth, I shall call a beable subalgebra of form $\mathcal{B}_{\{|v_{R_i}\rangle\}}$ (constructed in the manner just described) a *Bub-definite* subalgebra for the state $|v\rangle$. It is easy to see that a Bub-definite subalgebra for a state $|v\rangle$ will not be a subalgebra of $|v\rangle$'s definite set unless $|v\rangle$ is an eigenstate of R; but, when it is, the Bub-definite subalgebra will then *coincide* with the definite set of $|v\rangle$. Also Bub-definite subalgebras will not generally be commutative: the main exceptions are when H is two-dimensional, and when R is non-degenerate with all its eigenvalues attributed nonzero probability by $|v\rangle$.

6. Maximal beable subalgebras

Bub-definite subalgebras have the extra feature that their projections form a maximal determinate sublattice of the ortholattice of projections on H. This means that any enlargement of the projection lattice of a Bub-definite subalgebra $\mathcal{B}_{\{|v_{R_i}\rangle\}}$ generates an ortholattice from which it is impossible to recover the probabilities prescribed by $|v\rangle$ as a measure over the two-valued homomorphisms on the enlarged lattice.[37] Since I have stopped short of making any *a priori* assumptions about the lattice structure of the projections in Segalgebras, it is natural to ask whether Bub-definite subalgebras are still maximal *as Segalgebras*. Indeed, the general question of when (if ever) Segalgebras of beables are 'as big as one can possibly get' for a given state is philosophically interesting in its own right, since an answer would

[35] See also Bub & Clifton (1996).

[36] Bub's method turns out to include the Kochen–Dieks modal interpretation (discussed in Clifton 1995) as a special case – see Bub (1997), sect. 6.3.

[37] Bub (1997, sect. 4.3), Bub & Clifton (1996).

seem to set a limit on how far a simple realism of possessed values in quantum mechanics can be pushed.

Call a subalgebra $\mathcal{B}$ of $\mathcal{S}$ with beable status for a state ω a *maximal beable subalgebra* for ω if it is not properly contained in any other subalgebra of $\mathcal{S}$ with beable status for ω. An easy application of Zorn's lemma shows that a maximal beable subalgebra for any given state always exists. The following result gives an explicit (though not completely general) characterization of maximal beable subalgebras, covering the case of Bub-definite subalgebras.

Theorem 10 Let $\mathcal{S}(H)$ be the Segalgebra of all bounded self-adjoint operators on a Hilbert space H and let $|v\rangle$ be any state vector in H. Then if $|v_x\rangle$ $(x \in X)$ is any family of vectors in H satisfying:

1. the members of $\{|v_x\rangle\}$ are mutually orthonormal,
2. $|v\rangle$ is in the closed span of $\{|v_x\rangle\}$, and
3. $|v\rangle$ is not orthogonal to any member of $\{|v_x\rangle\}$,

$\mathcal{B}_{\{|v_x\rangle\}}$ is a maximal beable subalgebra of $\mathcal{S}(H)$ for the state $|v\rangle$. If H is finite-dimensional, the converse also holds, i.e. if $\mathcal{B} \subseteq \mathcal{S}(H_n)$ is a maximal beable subalgebra for $|v\rangle \in H_n$, then there exists a family of vectors $|v_x\rangle$ $(x \in X)$ in H_n satisfying 1.–3. such that $\mathcal{B} = \mathcal{B}_{\{|v_x\rangle\}}$ (and this family is unique up to phases).

Proof. For the first claim, assuming 2. the beable status of $\mathcal{B}_{\{|v_x\rangle\}}$ for $|v\rangle$ follows exactly as discussed for Bub-definite subalgebras at the end of the last section. So all that remains is to prove maximality. Note first that because the members of $\{|v_x\rangle\}$ are mutually orthogonal by 1., each is an eigenvector (with eigenvalue 0 or 1) of all the one-dimensional projection operators $P_{|v_x\rangle}$ $(x \in X)$. Consequently, the set $\{P_{|v_x\rangle}\}$ is contained in $\mathcal{B}_{\{|v_x\rangle\}}$. Now consider the subalgebra $\mathcal{T}$ generated by $\mathcal{B}_{\{|v_x\rangle\}}$ and any $A \notin \mathcal{B}_{\{|v_x\rangle\}}$. Our task is to show that $\mathcal{T}$ cannot have beable status for $|v\rangle$ — so suppose (for *reductio ad absurdum*) that $\mathcal{T}$ does. Then for any $x \in X$, $A \bullet P_{|v_x\rangle} \in \mathcal{T}$ and so beable status for $|v\rangle$ requires $A \bullet P_{|v_x\rangle} \in \mathcal{N}_{|v\rangle}$, i.e. $A \bullet P_{|v_x\rangle}|v\rangle = |0\rangle$ (thms 4, 7). Since $P_{|v_x\rangle}|v\rangle = c|v_x\rangle$ (with $c \neq 0$, by 3.), using the definition of the anti-symmetric product (eqn 6) yields

$$cA|v_x\rangle = P_{|v_x\rangle}(A|v\rangle) \ (= c'|v_x\rangle, \text{ for some } c') \tag{43}$$

which shows that A has $|v_x\rangle$ as an eigenvector. Since this is true for any $x \in X$, A lies in the definite sets of all the states $\{|v_x\rangle\}$, and therefore $A \in \mathcal{B}_{\{|v_x\rangle\}}$ contrary to hypothesis.

For the converse claim, suppose $\mathcal{B} \subseteq \mathcal{S}(H_n)$ is a maximal beable subalgebra for the state $|v\rangle \in H_n$. Consider the subspace of H_n given by

$$S = \{|w\rangle \in H_n : (A \bullet B)|w\rangle = |0\rangle \text{ for all } A, B \in \mathcal{B}\} \tag{44}$$

which is non-trivial since $\mathcal{B}$'s beable status for $|v\rangle$ requires that $|v\rangle \in S$. We first show that S is invariant under $\mathcal{B}$, i.e. $|w\rangle \in S$ implies $C|w\rangle \in S$ for any $C \in \mathcal{B}$. (In fact, for this part of the argument the dimension of the Hilbert space need not be finite.) To establish this, we need to show that if $|w\rangle \in S$, then $(A \bullet B)(C|w\rangle) = |0\rangle$ for any $A, B \in \mathcal{B}$. Using the fact that the C^* product of two operators X and Y is expressible as $XY = X \circ Y - iX \bullet Y$ (see the remarks following eqn 9), together with the supposition that antisymmetric products formed in $\mathcal{B}$ map $|w\rangle$ to zero and the definition of the symmetric product (eqn 3), one calculates

$$\begin{aligned} (A \bullet B)C|w\rangle &= ((A \bullet B) \circ C)|w\rangle - i((A \bullet B) \bullet C)|w\rangle && (45)\\ &= ((A \bullet B) \circ C)|w\rangle && (46)\\ &= 1/2(A \bullet B)C|w\rangle + 1/2C(A \bullet B)|w\rangle && (47)\\ &= 1/2(A \bullet B)C|w\rangle. && (48) \end{aligned}$$

But eqns 45 and 48 are consistent only if $(A \bullet B)(C|w\rangle) = |0\rangle$, as required.

Now, by the definition of S, all the operators in $\mathcal{B}$ commute on S. And since S is invariant under $\mathcal{B}$, restricting the action of any self-adjoint operator in $\mathcal{B}$ to S induces a self-adjoint operator on the subspace S. Since H_n – and thus S – is finite-dimensional, it follows by a well-known result that the operators in $\mathcal{B}$ share at least one complete set of common eigenvectors $\{|v_y\rangle\}$ ($y \in Y$) *on the subspace* S. The set $\{|v_y\rangle\}$ clearly satisfies 1., and also 2. since $|v\rangle$ lies in S. We can also arrange for 3. to be satisfied – while preserving satisfaction of 1. and 2. – by just dropping from the set $\{|v_y\rangle\}$ any vectors orthogonal to $|v\rangle$. So we can conclude, then, that there is at least one set of vectors $\{|v_x\rangle\}$ ($x \in X$) satisfying 1.–3. which are common eigenvectors of all the beables in $\mathcal{B}$. If so, then clearly $\mathcal{B} \subseteq \mathcal{B}_{\{|v_x\rangle\}}$, and the hypothesis that $\mathcal{B}$ is maximal for $|v\rangle$ delivers the required conclusion that $\mathcal{B} = \mathcal{B}_{\{|v_x\rangle\}}$ for some set satisfying 1.–3. (For uniqueness of this set, it is easy to see that if $\mathcal{B}_{\{|v_x\rangle\}} = \mathcal{B}_{\{|v_y\rangle\}}$ for two sets of vectors satisfying 1.–3., then those sets must in fact generate the same rays in H_n.) *QED*.

For finite-dimensional H, thm 10 yields a complete picture of maximal beable subalgebras for any vector state $|v\rangle$ on $\mathcal{S}(H)$: they simply correspond 1-1 with sets of vectors satisfying 1.–3. of the theorem, which then end up being common eigenvectors for all the elements of the algebra. If the set contains only a single vector, it must be $|v\rangle$ itself, and we obtain the orthodox subalgebra (i.e., $|v\rangle$'s definite set); if the set is an orthonormal basis for the Hilbert space, we obtain a commutative subalgebra; and if the set falls between these two extremes we obtain a subalgebra with identically the same structure as a Bub-definite subalgebra.

For infinite-dimensional H, the converse of thm 10 breaks down in the final paragraph of the proof at the point where the existence of a complete set of commuting eigenvectors on the finite-dimensional subspace S is invoked. For, obviously, if S could no longer be assumed finite-dimensional, some of the elements of $\mathcal{B}$ might then have no eigenvectors in S, much less any common ones. However, there is a way to reformulate the theorem so that it *does* generalize to the infinite-dimensional case – see Halvorson & Clifton (forthcoming, corollary 4.6) for details.

Thm 10 also generalizes in a different direction. The theorem only characterizes maximal beable status with respect to some given *vector state* $|v\rangle$ on $\mathcal{S}(H)$. But not every pure state on $\mathcal{S}(H)$ can be represented by a vector in H. For example, consider the sub-Segalgebra generated by any observable A with completely continuous spectrum (together with the identity). This subalgebra will be commutative, and (thus) possess dispersion-free states. Using a Hahn–Banach-type argument, one can show that any pure state of a sub-Segalgebra extends to a pure state on the whole algebra.[38] Thus each dispersion-free state $\langle\cdot\rangle$ of the subalgebra generated by A extends to a pure state ρ on $\mathcal{S}(H)$. But such a ρ could not possibly be represented by a vector $|v\rangle$ in H, otherwise the fact that ρ is dispersion-free on A would force $|v\rangle$ to be an eigenvector of A, when by hypothesis it has *no* eigenvectors! It follows that there are pure states on $\mathcal{S}(H)$ that qualify (in the algebraic sense) as states, but cannot be represented by any state vector in H.[39] For such cases, however, it is still possible to prove the following:

Theorem 11 The definite set of any pure state ρ on $\mathcal{S}(H)$ (whether or not it is a vector state) is a maximal beable subalgebra for ρ.

Proof. I need a non-trivial result due to Kadison & Singer (1959, thm 4). They show that the definite set of any pure state ρ on the $C^\star$-algebra $\mathcal{U}(H)$ of all bounded operators on a Hilbert space (which they define to be the set of all *self-adjoint* operators on H on which ρ is dispersion-free) is not properly contained in the definite set of any other state on $\mathcal{U}(H)$. It is not difficult to see that a state is pure on a Segalgebra $\mathcal{S}$ exactly when its extension to the

[38] See Segal (1947), lemma 2.2 or Kadison & Ringrose (1983), thm 4.3.13.
[39] See Segal (1947), corollary 4.2 or Kadison (1975), pp. 106–7.

$C^{\star}$-algebra generated by $\mathcal{S}$ is pure. So the Kadison–Singer maximality result also holds for the pure states on $\mathcal{S}(H)$, and in particular, those that are not vector states.

Now consider the definite set $\mathcal{D}_\rho$ of any pure state ρ on $\mathcal{S}(H)$. To show $\mathcal{D}_\rho$ is a maximal beable subalgebra for ρ, let Q denote the subalgebra generated by $\mathcal{D}_\rho$ together with any element $A \notin \mathcal{D}_\rho$. Note that $I - A \notin \mathcal{D}_\rho$, and either A or $I - A$ must lie outside of $\mathcal{N}_\rho$ (otherwise both would get value 0 in state ρ, yet those values must sum to 1). So there is a $B \in Q$ such that $B \notin \mathcal{D}_\rho$ and $B \notin \mathcal{N}_{\rho|_Q}$. Now suppose Q has beable status for ρ. Then thm 4 (part 2.) and thm 7 dictate that there is a dispersion-free state $\langle\cdot\rangle$ on Q such that $\langle B\rangle \neq 0$. It follows that $\langle\cdot\rangle$ will extend to a pure state on $\mathcal{S}(H)$ that has a definite set properly containing ρs – contradicting the Kadison–Singer maximality result. *QED.*

7. From quasicommutative to commutative beables

In this final section, I want to point out two ways of arguing for the commutativity of beables. The first way depends on rejecting the modal idea of letting a system's beables vary from one of its quantum states to the next, and the second way depends on the Reeh–Schlieder theorem of algebraic relativistic quantum field theory.

First, suppose one had reasons to demand that a quantum system possess a fixed set of beables for all its quantum states. For example, one could think that the idea of a physical magnitude having a definite value at one time but not at another is conceptually incoherent (though modalists would dispute this). Or one could think that the extra flexibility of having a state-dependent set of beables is not necessary; in particular, not needed to solve the measurement problem, since all measurement outcomes can be ensured simply by granting beable status to (essentially) a single observable, like position.[40]

Now it would be unreasonable to require a fixed set of beables to satisfy the statistics of absolutely *all* states, since they may not all be realizable in nature. An example is provided by state vectors in $L_2(\mathcal{R})$ that fail to lie in the domain of the momentum operator $-i\frac{\partial}{\partial x}$. Since $-i\frac{\partial}{\partial x}$ is only densely defined, state vectors lying outside its domain of definition do not dictate any well-defined expectation value for momentum, and might be thought

[40] This is a view Bell seems to have held, as can be seen in the second quotation I gave in sect. 1.

unphysical for that reason.[41] However, even without committing ourselves to definitive necessary and sufficient conditions for a state to count as physical, it seems reasonable to expect a system's physical states to make up a full set of states on the system's Segalgebra. (For example, it is easy to see that any dense set of the vector states in H defines a full set of states on $\mathcal{S}(H)$.[42]) Assuming this, we have the following converse to thm 5 that establishes commutativity for non-modal interpreters who seek a state-independent ontology of a system's categorical properties (as in Bohm's theory).

Theorem 12 If a subalgebra of a Segalgebra $\mathcal{S}$ has beable status for every state in a full set of states on $\mathcal{S}$, then it is commutative.

Proof. Let $\mathcal{B}$ be a subalgebra of $\mathcal{S}$ and Ω a full set of states on $\mathcal{S}$. If $\mathcal{B}$ has beable status for every state in Ω, thms 4 (part 3.) and 7 dictate that $\mathcal{B} \bullet \mathcal{B} \subseteq \mathcal{N}_{\omega|_\mathcal{B}} \subseteq \mathcal{N}_\omega$ for all $\omega \in \Omega$. But since Ω is a full set, it is easy to see that $\bigcap_{\omega\in\Omega} \mathcal{N}_\omega = \{0\}$, which forces $\mathcal{B}$ to be commutative. *QED.*

Thm 12 also allows us to diagnose exactly what goes wrong in Misra's (1967) argument against hidden-variables (in fulfilment of a promise I made at the end of section 3). Without distinguishing 'observables' whose measurement outcomes are determined by hidden-variables from those that correspond to *bona fide* beables of the system, Misra demands that the outcome of any measurement be determined by a dispersion-free state on the algebra of all *observables* of a system (1967, p. 856), which we have seen is only reasonable if all 'observables' are beables of the system. Misra also assumes that the hidden-variables must be adequate for recovering (after averaging) the expectation values of at least the physical states of a system, which he too assumes will be a full set. That this is a lethal combination of assumptions should be clear. If all 'observables' are treated as beables, and they are forced to satisfy the statistics of a full set of states of the system, then thm 12 dictates that in fact all the *observables* of the system would have to commute – which is absurd! Far from delivering a no-go theorem, this is simply one more confirmation of Bell's point that not all 'observables' can have beable status.

[41] A more sophisticated example is found in algebraic quantum field theory on curved spacetime, where the expectation value of the stress-energy tensor is not defined for all states of the field, but only for the so-called 'Hadamard' states; see Wald (1994, sect. 4.6).

[42] And for the Hadamard states referred to in the previous note, thm 2.1 of Fulling et al. (1981) establishes that they are dense in any Hilbert space representation of a globally hyperbolic spacetime's Segalgebra of observables.

A second way to argue that beables must be commutative arises out of the algebraic approach to relativistic quantum field theory. In that approach, one associates with each bounded open region O in Minkowski spacetime M a $C^\star$-algebra $\mathcal{U}(O)$ whose Segalgebra represents all observables measurable in region O. In the 'concrete' approach, the algebras $\{\mathcal{U}(O)\}_{O\subseteq M}$ are taken to be von Neumann algebras of operators acting on some common Hilbert space consisting of states of the entire field on M. If the collection of algebras $\{\mathcal{U}(O)\}_{O\subseteq M}$ satisfies a number of very general assumptions involving locality, covariance, etc. (the details of which need not detain us here[43]), then it becomes possible to prove the Reeh–Schlieder theorem, whose main consequence is that every state vector of the field with bounded energy is a *separating vector* for all the local algebras $\{\mathcal{U}(O)\}_{O\subseteq M}$. This means that no non-zero operator A in any local algebra $\mathcal{U}(O)$ can annihilate such a state vector.[44]

Now consider this result in the context of my analysis of beable subalgebras of the observables of a system. Here the role of the system is played by the quantum field in some bounded open region O of spacetime, and the question is which of the observables in $\mathcal{U}(O)$'s Segalgebra can be granted beable status. As we have seen, a subalgebra $\mathcal{B}(O)$ will have beable status for a state ω of the field if and only if $\mathcal{B}(O)$ is $\mathcal{N}_{\omega|_{\mathcal{B}(O)}}$-quasicommutative. But if ω corresponds to a state vector $|v\rangle$ with bounded energy, then since that vector is separating for $\mathcal{U}(O)$, we have

$$A \in \mathcal{N}_{\omega|_{\mathcal{B}(O)}} \Leftrightarrow \langle v|A^2|v\rangle = 0 \Leftrightarrow \langle Av|Av\rangle = 0 \Leftrightarrow A|v\rangle = |0\rangle \Leftrightarrow A = 0. \quad (49)$$

It immediately follows that:

Theorem 13 Subalgebras of local beables selected from the Segalgebras of local observables in relativistic quantum field theory must be commutative in any state of the field with bounded energy.

Notice that the orthodox approach to value definiteness reduces to absurdity in this context: since ω is dispersion-free on $A \in \mathcal{U}(O)$ exactly when

$$A - \omega(A)I \in \mathcal{N}_{\omega|_{\mathcal{B}(O)}} = \{0\}, \quad (50)$$

taking $\mathcal{B}(O)$ to be just the definite set of $\omega|_{\mathcal{B}(O)}$ yields only multiples of the identity operator as beables! Of course, there is still plenty of room left for a

[43] See Butterfield (1995).

[44] Haag (1992, sect. II.5.3); see also Redhead (1995b) for an elementary discussion.

simple realism of possessed values based on other more satisfactory sets of commuting local beables.

References

Bell, J. L. & R. Clifton (1995). 'QuasiBoolean Algebras and Simultaneously Definite Properties in Quantum Mechanics', *International Journal of Theoretical Physics* **34**, 2409–21.

Bell, J. S. (1987). *Speakable and Unspeakable in Quantum Mechanics* (Cambridge: Cambridge University Press).

Bohm, D. (1952). 'A Suggested Interpretation of the Quantum Theory in Terms of "Hidden Variables", Part II', *Physical Review* **85**, 180–93.

Bratteli, O. & D. W. Robinson (1987). *Operator Algebras and Quantum Statistical Mechanics, Vol I.* (Berlin: Springer-Verlag, 2nd edition).

Bub, J. (1997). *Interpreting the Quantum World* (Cambridge: Cambridge University Press).

Bub, J. & R. Clifton (1996). 'A Uniqueness Theorem for "No Collapse" Interpretations of Quantum Mechanics', *Studies in History and Philosophy of Modern Physics* **27**, 181–219.

Butterfield, J. N. (1995). 'Vacuum Correlations and Outcome Dependence in Algebraic Quantum Field Theory', in *Fundamental Problems in Quantum Theory*, ed. D. Greenberger and A. Zeilinger, *Annals of the New York Academy of Sciences* 755–68.

Clifton, R. (1995). 'Independently Motivating the Kochen–Dieks Modal Interpretation of Quantum Mechanics', *British Journal for Philosophy of Science* **46**, 33–57.

(1996). 'The Properties of Modal Interpretations of Quantum Mechanics', *The British Journal for Philosophy of Science* **47**, 371–98.

Dürr, D., S. Goldstein, & N. Zanghi (1996). 'Bohmian Mechanics as the Foundation of Quantum Mechanics', in J. T. Cushing *et al.* (eds), *Bohmian Mechanics and Quantum Theory: An Appraisal* (Dordrecht: Kluwer), pp. 21–44.

Fulling, S. A., F. J. Narcowic & R. M. Wald (1981). 'Singularity Structure of the Two-Point Function in Quantum Field Theory in Curved Spacetime, II', *Annals of Physics* **136**, 243–72.

Geroch, R. (1985). *Mathematical Physics* (Chicago: University of Chicago Press).

Haag, R. (1992). *Local Quantum Physics: Fields, Particles and Algebras* (Berlin: Springer-Verlag).

Halvorson, H. & R. Clifton (forthcoming). 'Maximal Beable Subalgebras of Quantum-mechanical observables'.

Heywood, P. & M. L. G. Redhead (1983). 'Nonlocality and the Kochen–Specker Paradox', *Foundations of Physics* **13**, 481–99.

Kadison, R. V. (1975). 'Operator Algebras', in J. H. Williamson (ed.), *Algebras in Analysis* (London: Academic Press), pp. 101–17.

Kadison, R. V. & J. R. Ringrose (1983). *Fundamentals of the Theory of Operator Algebras. Vol. 1* (London: Academic Press).

Kadison, R. V. & I. M. Singer (1959). 'Extensions of Pure States', *American Journal of Mathematics* **81**, 383–400.

Landsman, N. P. (1998). *Mathematical Topics between Classical and Quantum Mechanics* (New York: Springer-Verlag).

Mermin, N. D. (1993). 'Hidden Variables and the Two Theorems of John Bell', *Reviews of Modern Physics* **65**, 803–15.

Misra, B. (1967). 'When Can Hidden Variables be Excluded in Quantum Mechanics?', *Il Nuovo Cimento* **47A**, 841–59.

Pagonis, C. & R. Clifton (1995). 'Unremarkable Contextualism: Dispositions in the Bohm Theory', *Foundations of Physics* **25**, 281–96.

Plymen, R. J. (1968). 'Dispersion-Free Normal States', *Il Nuovo Cimento* **54**, 862–70.

Redhead, M. L. G. (1989). *Incompleteness, Nonlocality and Realism* (Oxford: Clarendon Press).

(1995a). *From Physics to Metaphysics* (Cambridge: Cambridge University Press).

(1995b). 'More Ado About Nothing', *Foundations of Physics* **25**, 123–37.

Rudin, W. (1974). *Real and Complex Analysis* (New York: McGraw-Hill, 2nd edition).

Segal, I. (1947). 'Postulates for General Quantum Mechanics', *Annals of Mathematics* **4**, 930–48.

Simmons, G. F. (1963). *Introduction to Topology and Modern Analysis* (New York: McGraw-Hill).

Van Fraassen, B. C. (1973). 'Semantic Analysis of Quantum Logic', in C. A. Hooker (ed.), *Contemporary Research in the Foundations and Philosophy of Quantum Theory* (Dordrecht: Reidel), pp. 80–113.

Wald, R. M. (1994). *Quantum Field Theory in Curved Spacetime and Black Hole Thermodynamics* (Chicago: University of Chicago Press).

Zimba, J. & R. Clifton (1998). 'Valuations on Functionally Closed Sets of Quantum Mechanical Observables and Von Neumann's "No-Hidden-Variables" Theorem', in P. Vermaas and D. Dieks (eds.), *The Modal Interpretation of Quantum Mechanics* (Dordrecht: Kluwer).

3

Aspects of objectivity in quantum mechanics

HARVEY R. BROWN

1. Introduction

Relative to some inertial coordinate system defined in Galilean spacetime, we assume that the quantum state of the physical system of interest satisfies the time-dependent Schrödinger equation

$$i\frac{d|\psi(t)\rangle}{dt} = H|\psi(t)\rangle \tag{1}$$

where the Hamiltonian H may itself be time-dependent. (Here, and throughout this paper, units are chosen so that $\hbar = 1$.) Suppose we require that eqn 1 be covariant under some (possibly time-dependent) unitary transformation represented by $|\psi(t)\rangle \to |\psi'(t')\rangle = U(t)|\psi(t)\rangle$ so that

$$i\frac{d|\psi'(t')\rangle}{dt'} = H'|\psi'(t')\rangle. \tag{2}$$

When $d/dt' = d/dt$, it can easily be shown that covariance holds if and only if

$$H' = UHU^{-1} + i\frac{\partial U}{\partial t}U^{-1}. \tag{3}$$

Recall that such covariance does not necessarily correspond to a symmetry of the Schrödinger dynamics: covariance of this general kind is expected to hold, for example, even for a wide class of non-linear coordinate transformations which are unitarily implementable. In such cases, the transformed Hamiltonian in eqn 3 will not not 'take the same form' as does H in eqn 1. In the case of a transformation to a rectilinearly accelerating

I wish to thank the referees for useful comments related to the original version of this essay, and Jeremy Butterfield for editorial improvements. I also thank Lucien Hardy, Peter Holland, Simon Saunders, Paul Teller and especially Guido Bacciagaluppi and Erik Sjöqvist for very helpful discussions, which prevented more misunderstandings than may be remaining.

coordinate system, say, the scalar potential in H' will contain a new term corresponding to the inertial force acting on the particle – so that in particular a free particle no longer 'looks' free.

Something like the opposite of this familiar process can also occur. In very special cases of background potentials, such as that of a time-dependent simple harmonic potential, a quantum particle can exhibit *free* motion when described relative to an appropriately accelerating coordinate system.[1] In transforming in this case from an inertial coordinate system (with respect to which the potential is defined) to such a contrived non-inertial coordinate system, the energy spectrum of the particle goes from being purely discrete to purely continuous – a surprising state of affairs perhaps but consistent with non-invariance of the Hamiltonian as seen in eqn 3 above. Even the existence of tunnelling in some cases turns out to be coordinate-dependent.[2]

Does this mean that it is not always an objective state of affairs as to whether a given particle is free, or whether tunnelling is taking place? This is similar to the question as to whether the Newtonian forces acting on a specific classical particle are not objective, given the ability to transform to the rest frame of the particle – relative to which obviously it 'moves freely'. The answer to both these questions is surely negative. To me, at any rate, the 'freedom' of the above-mentioned quantum particle when described relative to a contrived accelerating frame, for example seems just as much of an artifact as the background thermal radiation 'seen' – via the Unruh effect – by a uniformly accelerating detector in the (inertially-defined) vacuum field in Minkowski spacetime. There is no doubt more to be said about this issue, but it will be skirted here. Questions of objectivity – in so far as they are concerned with the issue of coordinate-independence in Galilean spacetime – will be restricted in the paper to the context of differing inertial perspectives.

Section 2 contains a review of the covariance of the Schrödinger equation under local gauge transformations and Galilean coordinate transformations. In the subsection on gauge covariance, I include a brief discussion of the sense in which the gauge principle can be said to 'generate' dynamical gauge fields (such as the Maxwell field in the case of the $U(1)$ symmetry). The motivation of this discussion is two-fold. In part the discussion attempts to address the 'mystery' associated with the gauging procedure, recently highlighted by Teller. It also touches on the connection between gauge

[1] See Kuchař (1980), sects VII and VIII, and the independent work of Takagi (1990).
[2] See Takagi (1991).

covariance in quantum theory and the requirement of general covariance in the general theory of relativity.

Before summarizing the content of the subsequent sections, a few more introductory words are in order. Note that even in the case of the symmetries discussed below in section 2, the second term on the RHS of eqn 3 is generally of importance. Eqn 3 forces us to be wary in accepting the common claim that under a unitary transformation U implementing a symmetry, a self-adjoint operator A representing some physical magnitude itself transforms unitarily as $A \to A' = UAU^{-1}$. Such a transformation ensures of course that the mean value of A' (or rather its associated physical magnitude) defined with respect to the transformed state equals that of A defined with respect to the original state. But in the case, say, of a Galilean coordinate transformation – where the associated unitary transformation of the state is time-dependent – eqn 3 is telling us that UHU^{-1} does not represent the same magnitude relative to the 'moving' observer as H does relative to the 'stationary' one. Indeed, one would not expect the total energy to be invariant under a passive Galilean boost, and it is far from being the only interesting magnitude for which the corresponding operator fails to transform unitarily in such cases as coordinate and gauge transformations.[3]

This state of affairs implies that a system undergoing Hamiltonian evolution which describes a closed loop in ray space (projective Hilbert space) relative to a given inertial coordinate system will generally fail to preserve this closure property when viewed from a (spatially) translated coordinate system if the translation is time-dependent, and in particular when viewed from a boosted coordinate system. (A special case is the non-preservation of stationarity.) The ensuing issue of the Galilean coordinate-dependence of the geometric phase defined on paths in ray space will be taken up in section 3 below. Some commentators have regarded such coordinate dependence as a defect in the standard formulation of geometric phase, but I am not so sure.

The existence of non-unitary, symmetry-related transformations of operators also raises a question as to the objectivity of 'sharp values' which a number of related interpretations of quantum mechanics attribute, *qua* elements of reality, to certain magnitudes of a system under certain conditions. In particular, the question as to whether sharpness imposed by the eigenstate–eigenvalue link is invariant under coordinate transformations and/or gauge transformations is briefly examined in section 4.

[3] For a discussion of the sense in which the 'common claim' above regarding the transformation $A \to A' = UAU^{-1}$ is correct, see Brown & Holland (1999) and particularly Brown, Suàrez & Bacciagaluppi (1998).

In the final section of the essay, attention is turned to the theory of quantum reference frames, due primarily to Aharonov and Kauffher. It would seem that this theory reformulates and reinforces the lesson urged in section 4, namely that sharp values of observable magnitudes in quantum mechanics must, in a rather special (non-classical) sense, be given a relational, and not an absolute status. This, at least, is the conclusion of section 5.

The overlapping issues of symmetry and objectivity in quantum mechanics taken up in this paper – which is largely based on a number of recent collaborative efforts – reflect prominent themes in the work of Michael Redhead. Michael's influence on me over the last quarter-century as teacher, mentor, collaborator and friend has been enormous. It is with pleasure and gratitude that I dedicate the essay to him.

2. Covariance of the Schrödinger equation

2.1 Gauge transformations

Consider the Schrödinger equation for a single, spinless particle evolving in the presence of a vector potential $\mathbf{A}(\mathbf{x}, t)$ and scalar potential $V(\mathbf{x}, t)$, which may or may not be of electromagnetic origin:[4]

$$i\frac{\partial\psi(\mathbf{x}, t)}{\partial t} = \left[\frac{-1}{2m}\{\nabla - i\mathbf{A}(\mathbf{x}, t)\}^2 + V(\mathbf{x}, t)\right]\psi(\mathbf{x}, t) \tag{4}$$

It is sometimes claimed that the form of the Hamiltonian in eqn 4 is itself a direct consequence of Galilean invariance (or rather Galilean kinematics), a claim we shall briefly return to later.[5] In the meantime, let me repeat the well-known fact that eqn 4 is covariant under a local gauge transformation $\psi(\mathbf{x}, t) \to \psi'(\mathbf{x}, t) = \exp[i\xi(\mathbf{x}, t)]\psi(\mathbf{x}, t)$ when the potentials transform as:

$$\left.\begin{aligned} \mathbf{A}'(\mathbf{x}, t) &= \mathbf{A}(\mathbf{x}, t) + \nabla\xi(\mathbf{x}, t) \\ V'(\mathbf{x}, t) &= V(\mathbf{x}, t) - \frac{\partial\xi(\mathbf{x}, t)}{\partial t} \end{aligned}\right\} \tag{5}$$

In the case of Maxwell fields (where both eqn 4 and the transformation of the wavefunction above hold in the case of a particle with unit charge), the

[4] An example of the appearance of a vector potential which is not related to an external magnetic field is mentioned in section 3.5 below.

[5] See note 40 below.

transformations 5 leave electric and magnetic field strengths unaltered – even in the non-relativistic limit as we shall see in the next subsection.

It is well-known too that the geometric role of the vector potential **A** is that of determining a connection, associated with the 'covariant' derivative $\mathbf{D} \equiv \nabla - i\mathbf{A}(\mathbf{x})$.[6] We can thus define the gauge-invariant quantity:

$$f_{kl} \equiv i[D_k, D_l] = \frac{\partial A_l}{\partial x_k} - \frac{\partial A_k}{\partial x_l} \tag{6}$$

where $k, l = 1, 2, 3$, which can be called the curvature tensor associated with the connection determined by **A**. In the case where **A** has electromagnetic origin, f_{kl} is non-zero wherever there exists a magnetic field – and it is detectable in quantum mechanics even when the particle is entirely excluded from the region where the magnetic field is confined, as was famously shown by Aharonov and Bohm in 1959. This result demonstrated the importance of interpreting the electromagnetic field as a gauge field corresponding to the local $U(1)$ group (see below).

Perhaps it is worth noting here that the relationship between the gauge principle and the introduction of a rule of parallel transport, and hence a connection, has also been usefully exploited for fibre bundles where the base space is other than physical space (or spacetime). In particular, it has been known for some time that the requirement of symmetrization and antisymmetrization of the states of collections of identical particles in quantum mechanics is intimately linked with the global topological structure of the reduced configuration space defined for N particles moving in a three-dimensional physical space.[7] (This reduced space is obtained by identifying points in the standard $3N$-dimensional configuration space which are related

[6] This geometric view of the potential is usually discussed in relation to the 4-potential in relativistic quantum mechanics, to which we turn shortly. In case it needs stressing, the connection associated with this 4-potential, and the curvature it induces, are defined relative to the $U(1)$ fibre bundle whose base space is spacetime and whose fibres are the 'internal spaces' related to the local phases of the complex wavefunction. If one wants to say that such curvature is not a genuine curvature of spacetime, unlike that defined in general relativity, the reason can only be that the affine connection coefficients and the associated curvature tensor in the latter theory are defined with respect to the tangent bundle of spacetime. (Note that strictly speaking, in the case of the electromagnetic potential in quantum theory, both the covariant derivative operator and the connection coefficients actually depend on the charge of the particle as well as the 4-potential, so the connection depends on the type of particle involved. But the curvature operator – see eqn 8 below – is defined to be charge-independent; see Lawrie (1990), §8.1, for a nice introductory treatment of these issues.)

[7] See in particular Leinaas & Myrheim (1977). For a brief review of this approach, and its particular naturalness in the context of de Broglie–Bohm pilot-wave theory, see Brown, Sjöqvist & Bacciagaluppi (1998).

by permutation of particle labels, and removing the singular points corresponding to spatial coincidence of two or more particles.) Suppose one introduces quadratically integrable wavefunctions subject to a local gauge symmetry on the reduced configuration space, which itself turns out to be doubly connected. Then it can be shown that the wavefunction must either change sign or remain invariant under an exchange of the particles.[8] Furthermore, the former fermionic case is somewhat analogous to the force-free Aharonov–Bohm effect, but this time the 'gauge field' is confined to the off-limits singularities in the reduced configuration space. Now it may be open to interpretation just how explanatory this topological view of fermionic statistics is, but the overall approach seems to offer a deep insight into the range of possible statistics the particles may in principle exhibit.[9] The point I want to stress is that the local gauge principle on the reduced configuration space is central to the argument.

Returning to the familiar gauge fields on spacetime, I wish finally to comment on a claim sometimes made in the context of the relativistic quantum mechanics. This is the claim that the gauge field (electromagnetism) *has* to be introduced to ensure covariance of the equation of motion under local $U(1)$ gauge transformations.[10] Now something almost magical *seems* to be occurring here, and one can sympathize with the puzzlement recently expressed by Teller (1997) at the apparent fact that in gauge theories generally, a change in mere 'conventions' (local gauge) can have 'dramatic repercussions in seeming to force the introduction of an otherwise neglected physical field!' The remaining comments in this subsection are offered in the hope of finding some alleviation of this puzzlement.

Let us consider then the case of a free spin-$^1/_2$ particle. As is well-known, the Dirac equation (in natural units wherein $c = \hbar = 1$)

$$(i\gamma^{\mu}\partial_{\mu} - m)\psi(x) = 0 \tag{7}$$

'becomes' gauge covariant under $\psi(x) \to \psi'(x) = \exp[i\lambda\xi(x)]\psi(x)$ if the ordinary derivative ∂_{μ} acting on the spinorial wavefunction ψ is replaced by the gauge-coviariant derivative $D_{\mu} = \partial_{\mu} + i\lambda A_{\mu}(x)$ – resulting in a relativistic analogue of eqn 4. Here λ is a 'coupling' constant (ultimately related to the charge of the particle, but at this point we can understand it merely as

[8] The restriction to these two options does not hold if physical space is less than three-dimensional, allowing for the possibility of anyons in the two-dimensional case.

[9] More recently, a derivation of the spin-statistics relation has resulted from the construction, within the same approach, of an exchange operator for identical particles with spin, with further topological significance in an enlarged Hilbert space; see Berry & Robbins (1997).

[10] See, for example, Ryder (1987), p. 99.

a factor such that A_μ need not have the same units as ∂_μ), and the 4-vector field A_μ determines a connection on the $U(1)$ fibre bundle whose base manifold is spacetime. This vector field transforms as $A'_\mu(x) = A_\mu(x) - \partial_\mu \xi(x)$. The very term 'minimal coupling' standardly associated with this procedure reflects the fact that the resulting equation of motion for the Dirac wavefunction is not the only conceivable gauge covariant equation which reduces to the original eqn 7 when $A_\mu = 0$, but it is the simplest.

But why should we be interested in a gauge-covariant version of eqn 7 in the first place? Following Nakahara, we might relate the gauge principle to the truism that '*physics should not depend on how we describe it*', and compare it with the requirement of general covariance in the general theory of relativity.[11] Yet is it *a priori* obvious that the choice of local phase of the wavefunction is merely a choice of description? Hardly. (I remain to be convinced even that the italicized claim above constitutes the decisive justification of general covariance in general relativity, but that's another story.) Surely it is hindsight, provided by the link with electromagnetism, that makes it appear so.[12] An entirely different issue, however, is the significance of introducing a connection, and hence a rule of parallel transport in the $U(1)$ bundle. It is this which tells us how to compare distant phases of the wavefunction, a natural requirement which does not, in itself, beg the question of gauge covariance.

What has actually been done above in replacing the ordinary derivative in eqn 7 by the (gauge) covariant derivative is, however, analogous to the coordinate-general reformulation, in a *flat* affine spacetime, of the geodesic equation – essentially the equation of motion for a free classical particle – originally expressed in component form relative to a global inertial coordinate system (i.e., one in which the connection coefficients vanish everywhere). Again, this reformulation is carried out effectively by replacing the ordinary partial derivative by the covariant derivative associated with the affine connection defined on the tangent bundle. (For the sake of completeness, the details are given in the appendix.)

In this case, the generally covariant equation of free motion captures the effect of 'inertial forces' acting on the particle from the perspective of a non-inertial coordinate system. In the case of the gauge covariant version of the free Dirac equation, one could perhaps say that the appearance of the 'potential' A_μ in the Dirac Hamiltonian similarly captures the effect of

[11] Nakahara (1990), p. 10.

[12] A related issue is why certain quantities in physics are gauged and not others (recently raised by Teller (1997), p. 517).

'pure-gauge forces' arising out of the generalization to arbitrary gauges. But *in neither case* is the connection a *bona fide* dynamical object yet nor are the mentioned 'forces' of dynamical origin. This only comes about when (i) the connection gives rise to curvature, and (ii) when the matter field acts back on the connection.

Condition (i) is of course *inconsistent* with eqn 7, in that there is no gauge in which $A_\mu = 0$ everywhere (just as there are no global inertial coordinates systems when the curvature tensor in the spacetime tangent bundle is non-zero).[13] The gauge-invariant curvature is defined in terms of the commutator of the components of the covariant derivative (a generalization of eqn 6):

$$F_{\mu\nu} \equiv -\frac{i}{\lambda}[D_\mu, D_\nu] = \partial_\mu A_\nu - \partial_\nu A_\mu. \tag{8}$$

Condition (ii) requires the introduction of an analogue of Einstein's field equations, which would determine *inter alia* the effect of the Dirac particle on the gauge potential A_μ. Now if we view eqn 7 as arising from an action principle, then a gauge-invariant action is already obtained just by replacing the ordinary derivative in the action associated with free motion by the gauge-covariant derivative. What is still lacking is a further gauge-invariant contribution to the action responsible for the back-action of the particle on the potential. The simplest Lorentz scalar that can be constructed from the gauge-invariant curvature is $F_{\mu\nu}F^{\mu\nu}$, and if, as is well-known, a term proportional to this scalar is added to the action, equations of motion for the gauge field are obtained which take the same form as Maxwell's equations when the electric current density $j^\mu(x)$ is interpreted as proportional to $\bar{\psi}(x)\gamma^\mu\psi(x)$, and the coupling constant λ is related to electric charge.[14]

It is striking that in this important and suggestive 'derivation' of (classical) electrodynamics Nature seems to oblige in the sense of adhering both to the minimal coupling procedure[15] and the choice of the simplest gauge- and Lorentz-covariant contribution to the action that accounts for the dynamical properties of the gauge field. But the point I wish to stress is a different one: *neither conditions (i) nor (ii) above are consequences of the requirement*

[13] Once one introduces curvature along with minimal coupling, a weaker claim than Ryder's above can reasonably be made: 'the electromagnetic properties of elementary fermions can be deduced simply by demanding that the Lagrangian be invariant under local phase transformations' (Collins et al. 1989, p. 48). The argument here requires identifying the gauge potential with the *known* electromagnetic potential, and λ with the charge of the fermion; the energy of the electromagnetic field is taken as given (Collins et al. 1989, p. 49).

[14] See, for example, Lawrie (1990), p. 143.

[15] For a discussion of the electromagnetic significance of minimal coupling, and some subtleties in its definition that have been overlooked here, see Sakurai (1964), pp. 182–3.

of local gauge covariance. Condition (i) is motivated by physical phenomena like that of Aharonov and Bohm, which indicate that certain observable interference effects in quantum theory can be interpreted as anholonomies associated with the curvature of the gauge connection.[16] But the existence of such curvature – and hence of the electromagnetic field – is not strictly required by the process of constructing a gauge-covariant reformulation of eqn 7.[17] Condition (ii) is motivated by the action–reaction principle holding between matter and the gauge field.[18] Its analogue in general relativity is the requirement of the non-existence of 'absolute objects' defined on the spacetime manifold (which are taken to act on the matter fields but are not acted upon), or what Wald[19] has somewhat misleadingly called the 'principle of general covariance'. This principle is far stronger than the mere requirement that the field equations be written in general-covariant form (i.e. expressed in the tensor calculus),[20] which is the tangent bundle analogue of gauge covariance.

These remarks should not be construed as disparaging the heuristic importance of treating electromagnetism as a gauge theory, which is of course illustrated in the successful use of the gauge principle in the standard model of particle physics. The postulation of physical gauge fields such as the gluon field in chromodynamics was made possible by the earlier recognition of the gauge structure of the electromagnetic field, and the pioneering investigation in 1954 of possible gauge fields associated with the $SU(2)$ group by Yang and Mills. But as before, direct observable effects of the gauge potentials in question can in principle be expected only under the assumption that the gauge fields – the generalizations of eqn 8 for non-Abelian gauge groups – are non-vanishing and satisfy the action–reaction principle.

We shall return to the issue of gauge invariance and some of its implications in quantum mechanics in section 4.

[16] Although the Aharonov–Bohm (AB) effect was instrumental in the recognition of electromagnetism as a gauge field in quantum theory, the consideration here does not depend on the gauge curvature being inaccessible to the quantum system as in the AB effect.

[17] Note that Rai Dastidar & Rai Dastidar (1994, 1995) have already argued that local gauge invariance in quantum theory does not imply the existence of an external electromagnetic field, but they do not (I think) make it clear that the gauge potentials introduced into their gauge invariant formulation of the quantum dynamics in the case of free particles correspond to a flat connection.

[18] See Lawrie (1990), p. 161, and Anandan (1997), sect. 5.4. For a wider discussion of the role of the action–reaction principle in modern physics, see Anandan & Brown (1995).

[19] Wald (1984), p. 57.

[20] For a fuller discussion of this point, see Brown & Sypel (1995), sect. 4.

2.2 Galilean covariance[21]

Let us consider a passive coordinate transformation from the frame F (relative to which eqn 4 holds) to the frame F' moving at uniform velocity $\mathbf{v}$ relative to F:

$$\mathbf{x}' = \mathbf{x} - \mathbf{v}t, \quad t' = t. \tag{9}$$

Relative to F', the value of the wavefunction at an arbitrary spacetime location is related to that of ψ at the same location by a phase factor

$$\psi'(\mathbf{x}', t') = e^{i\phi}\psi(\mathbf{x}, t), \tag{10}$$

in order to ensure invariance of the probability density at that location. Textbook treatments[22] of the Galilean covariance of the Schrödinger equation invariably deal with the special cases of a free particle or one with finite scalar potential V, where it is shown that covariance is secured, assuming V transforms as a scalar field, when

$$\phi = m(\mathbf{v}^2 t/2 - \mathbf{v}.\mathbf{x}). \tag{11}$$

However, since we do not expect the phase in eqn 10 to depend on the dynamics, we expect it to be independent of the vector potential $\mathbf{A}$, as well as V. Indeed, it can easily be shown[23] that eqn 4 is Galilean-covariant under eqns 9, 10 and 11 when the potentials transform as

$$\left.\begin{aligned} \mathbf{A}'(\mathbf{x}', t') &= \mathbf{A}(\mathbf{x}, t) \\ V'(\mathbf{x}', t') &= V(\mathbf{x}, t) - \mathbf{v}.\mathbf{A}(\mathbf{x}, t) \end{aligned}\right\} \tag{12}$$

(In fact eqns 11 and 12 are necessary conditions for the covariance of eqn 4, ignoring irrelevant gauge transformations and arbitrary constants in the phase eqn 11.) We see that in the general case, the scalar potential V no longer transforms as a scalar field.

To see how eqn 11 involves a gauge transformation[24] – although as is stressed at the end of this subsection, a Galilean transformation is more than just a gauge transformation – let us briefly consider what happens when we

[21] This subsection contains a brief summary of the detailed review of the Galilean covariance of quantum mechanics given in Brown & Holland (1999).

[22] See, for example, Ballentine (1990, sect. 4.3).

[23] See Takagi (1991, p. 465) and Brown & Holland (1999). It is perhaps surprising that the demonstration of covariance in the general case (involving the vector as well as scalar potential) is so rare in the literature.

[24] The following argument is taken from Takagi (1991, p. 465), where the treatment involves the more general case of a time-dependent $\mathbf{v}$.

suppose that the wavefunction transforms like a scalar field, which we write as $\psi(\mathbf{x}, t) \to \varphi'(\mathbf{x}', t') = \psi(\mathbf{x}, t)$. In this case, still using eqn 12, one obtains

$$i\frac{\partial \varphi'}{\partial t'} = \left[\frac{-1}{2m}\{\nabla' - i(\mathbf{A}' + m\mathbf{v})\}^2 + V' - m\mathbf{v}^2/2\right]\varphi'. \tag{13}$$

Here we see that in relation to eqn 4, the transformed Hamiltonian picks up a curl-free – and hence uninteresting – vector potential $m\mathbf{v}$ (itself a case of a flat connection), as well as the extra constant term $-m\mathbf{v}^2/2$. It is largely a matter of convenience whether we absorb this latter term into V', but at any rate both terms can now be eliminated by the appropriate gauge transformation of φ':

$$\left.\begin{aligned}\psi'(\mathbf{x}', t') &\equiv \exp[im(-\mathbf{v}.\mathbf{x}' - \mathbf{v}^2 t'/2)]\varphi'(\mathbf{x}', t') \\ &= \exp[im(-\mathbf{v}.\mathbf{x} + \mathbf{v}^2 t/2)]\psi(\mathbf{x}, t)\end{aligned}\right\} \tag{14}$$

thus recovering the phase in eqn 11 and ensuring the covariant form of the wave-equation for $\psi'(\mathbf{x}', t')$.[25]

It is worth noting that one can find a set of electromagnetic 'field' equations, which can be considered the non-relativistic limit of Maxwell's equations in the case where magnetic effects predominate over electric ones, which are strictly Galilean covariant under (12).[26] (The field equations are identical to Maxwell's except for the absence of the displacement current: time-varying electric fields do not induce a magnetic field. The result is a phenomenological theory of magnetostatics, corresponding to the usual macroscopic situation where negative and positive charges cancel.) The electric and magnetic fields are related to the scalar and vector potentials in the usual way. Hence a fully Galilean-covariant quantum theory of the Schrödinger field interacting with an external electromagnetic field is possible. It should however be emphasized here that a necessary condition for the covariance of this theory is that the Schrödinger current and charge density do not act as sources of the Maxwell field.

[25] Note that eqn 11 ensures the covariance of the fundamental de Broglie relation between momentum and wavelength in the case of plane wave solutions of eqn 4 when the potentials are identically zero. The fact that wavelength, like momentum, does not transform invariantly under a Galilean transformation indicates that the wavefunction is quite different from a classical wave amplitude, a fact which is frequently forgotten when appeal is made in quantum mechanics to the Bohrian complementarity between the 'classical' pictures of wave and particle. For further details, see Lévy-Leblond (1976).

[26] These field equations correspond to the 'magnetic' (non-relativistic) limit of Maxwell's equations formulated in Le Bellac & Lévy-Leblond (1973). It is shown in this work that there are two natural nonrelativistic limits, which makes it difficult to say which electrodynamic effects are strictly relativistic in nature.

The results reviewed so far allow us to analyse the problem of covariance in the de Broglie–Bohm (de B–B) pilot-wave interpretation of quantum mechanics. Recall that the guidance equation for the de B–B corpuscle takes the following form when the system is in the presence of a vector potential $\mathbf{A}$:[27]

$$m\dot{\mathbf{x}} = \nabla\mathcal{S} - \mathbf{A} \tag{15}$$

where $\mathcal{S}$ is the phase of the 'guiding field': $\psi(\mathbf{x}, t) = R(\mathbf{x}, t)\exp[i\mathcal{S}(\mathbf{x}, t)]$. (Again, if $\mathbf{A}$ is of electromagnetic origin, eqn 15 holds when the particle has unit charge.) Here, as in eqn 4, the role of the vector potential can be seen as a compensating term: it renders the velocity of the corpuscle invariant under local gauge transformations, as the reader can easily check.

But is eqn 15 consistent with Galilean kinematics? From eqns 11 and 12 we see that

$$\begin{aligned} m\dot{\mathbf{x}}' &= \nabla'\mathcal{S}' - \mathbf{A}' \\ &= \nabla\mathcal{S} - m\mathbf{v} - \mathbf{A} \\ &= m\dot{\mathbf{x}} - m\mathbf{v}, \end{aligned} \tag{16}$$

so we recover the Galilean transformation of a momentum vector. Furthermore, it can be shown from eqn 12 that the classical Lorentz force acting on the de B–B corpuscle when immersed in an electromagnetic field (which exists as well as the force due to the Bohm quantum potential – itself also being affected by the external field) transforms invariantly, as one expects of a classical force in Galilean space-time.

This state of affairs might appear comforting. But for those who regard the true de B–B dynamics as captured in the first-order eqn 15, the forces acting on the corpuscle and generated by the guiding wave are Aristotelian, not Newtonian: they produce velocities not accelerations. This entails that there is a natural state of 'motion', which is rest. Yet no privileged frame is picked out by the 'hidden' dynamics of the corpuscle, and what may at first sight have looked like a sign of strength of the theory is now seen as a possible source of embarrassment.[28]

We finish this brief review of Galilean covariance with a glance at the state of affairs in the abstract formalism, since we will refer to some of the details later. Here we replace eqn 4 by

[27] See Holland (1993).

[28] For further discussion of this issue, see Brown et al. (1996) and Valentini (1997).

$$i\frac{d|\psi(t)\rangle}{dt} = \left[\frac{\{\mathbf{P} - \mathbf{A}(\mathbf{Q})\}^2}{2m} + V(\mathbf{Q}, t)\right]|\psi(t)\rangle, \tag{17}$$

where **P** is the operator representing the canonical momentum and **Q** is again the position operator. (We omit putting hats on symbols representing operators.) Eqn 17 is covariant under a passive Galilean boost implemented by a unitary transformation, represented by $|\psi(t)\rangle \rightarrow |\psi'(t')\rangle = U_G|\psi(t)\rangle$ when **P** and **Q** transform invariantly – note, not unitarily – and

$$\left.\begin{aligned} \mathbf{A}'(\mathbf{Q}', t') &= U_G[\mathbf{A}(\mathbf{Q}, t)]U_G^{-1} \\ V'(\mathbf{Q}', t') &= U_G[V(\mathbf{Q}, t) - \mathbf{v}.\mathbf{A}(\mathbf{Q}, t)]U_G^{-1} \end{aligned}\right\} \tag{18}$$

and when

$$U_G = \exp(im\mathbf{v}^2 t/2)\exp(i\mathbf{v}.\mathbf{P}t)\exp(-im\mathbf{v}.\mathbf{Q}). \tag{19}$$

The form of eqn 19[29] should come as no surprise, given eqn 11. The 'extra' factor in eqn 19 containing **P** simply accounts for the fact that a passive Galilean transformation involves a time-dependent translation of the coordinate axes (as well as a velocity transformation). This is already taken into account in eqn 10 where we are comparing the values of the primed and unprimed wavefunctions at the same space-time location, or in other words for different arguments (coordinates) – which also accounts for the absence of the unitary factors in eqn 12 when compared with eqn 18. (It is this feature that makes a passive boost something beyond a mere gauge transformation.) The unitary operator implementing an active boost of the particle plus potentials is given by the inverse of that of eqn 19, and is of course consistent with the role of the canonical operators $-\mathbf{P}$ and **Q** as generators in quantum mechanics of active translations and (instantaneous) boosts respectively.

3. The non-invariance of geometric phase

Following the seminal work of Berry (1984) on systems undergoing cyclic adiabatic evolution, it has come to be realized that there exists an important

[29] See also Fonda & Ghirardi (1970) sect. 2.5. Note that eqn 19 is independent of the dynamics: the potentials in the Hamiltonian in eqn 17 make no appearance. This is to be expected given the properties of Galilean space-time. However, the same situation will not hold for the analogue of eqn 19 in relativistic quantum mechanics given the relativity of simultaneity in Minkowski spacetime.

geometrical structure in the quantum formalism related to the phase of a quantum system undergoing Schrödinger evolution. Let us consider a system undergoing cyclic (not necessarily adiabatic) evolution, so that during the temporal interval $[0, T]$, the system's final and initial states coincide up to a phase factor: $|\psi(T)\rangle = \exp(i\phi)|\psi(0)\rangle$, where ϕ is an arbitrary real number. When projected onto ray space, i.e. the projective Hilbert space $\mathscr{P}$, this evolution defines a closed path. Now suppose we have the idea of subtracting from the total phase ϕ the accumulation of local phase changes produced by the motion on this path. By 'a local phase change' is meant the quantity $\delta\phi(\psi_t, \psi_{t+\delta t}) = -i\langle\psi(t)|d/dt|\psi(t)\rangle\delta t$. We are subtracting then from the total phase the quantity

$$\left.\begin{aligned} \phi_d &\equiv -i\int_0^T \langle\psi(t)|d/dt|\psi(t)\rangle dt \\ &= -\int_0^T \langle\psi(t)|H|\psi(t)\rangle dt \end{aligned}\right\} \tag{20}$$

where H is again the Hamiltonian responsible for the cyclic motion (still putting $\hbar = 1$). Because it depends on H, the quantity ϕ_d is called the dynamical phase. Now what we are left with after the subtraction, $\phi_g = \phi - \phi_d$, is the 'geometric phase', formulated by Aharonov & Anandan (1987). It is reparametrization invariant (i.e. independent of the speed at which the path in $\mathscr{P}$ is traversed). Moreover, it takes the same value for all the (infinity of) evolutions in the Hilbert space which project onto the given closed path in $\mathscr{P}$; it is a property only of that path. It is natural then to interpret it as the anholonomy associated with 'parallel transport' – transport in which there is no local phase change – around the closed curve in $\mathscr{P}$. The existence of geometric phase testifies to the existence of a non-flat connection on $\mathscr{P}$.

I shall return at the end of this section to the significance of the discovery of the curved geometry of the ray space in quantum mechanics. At this point I wish to mention a recent result concerning the Galilean non-invariance of geometric phase.

A curve in spacetime is a geometrical object; in particular whether it is closed or not does not depend on the choice of coordinate system. Now imagine a perfectly elastic ball bouncing on the hard floor of a laboratory; to an observer at rest relative to the lab, the evolution of the ball defines a closed path in state (phase) space. Yet to an observer in uniform motion relative to the lab, it doesn't: the ball does not return to the same spatial location at each bounce.[30] Note that it is not the speed *per se* of the moving observer that counts here, but rather the fact that the new frame is spatially translated in a time-dependent way relative to the lab frame. Analogously, we should not expect in quantum mechanics that the closure property of curves in $\mathscr{P}$ should be invariant under time-dependent unitary transformations of the state. In particular, we should expect, and indeed it is so, that closure is (inertial-) frame-dependent, given the time-dependence of the second exponent in eqn 19 – which, as noted above, has to do with passive time-dependent translations and not (instantaneous) boosts *per se*.

The fact that the very condition for the definability of geometric phase (closure of the path in $\mathscr{P}$) is not generally preserved under time-dependent unitary transformations (including gauge transformations) was recognized from the start. Indeed, it was recognized that even in the special cases (which exclude unitary implementations of Galilean transformations) where closure is preserved, the geometric phase is still not invariant.[31] However, an interesting development – particularly from the point of view of Galilean transformations – had to do with the formulation of a geometric phase factor γ_g for open paths in $\mathscr{P}$ given independently by Aitchison & Wanelik (1992) and Mukunda & Simon (1993), and which is defined by

$$\exp(i\gamma_g) = \left(\frac{\langle\psi(0)|\psi(T)\rangle}{\langle\psi(T)|\psi(0)\rangle}\right)^{1/2} \exp\left(-\int_0^T \langle\psi(t)|H|\psi(t)\rangle dt\right) \tag{21}$$

for evolutions such that $\langle\psi(T)|\psi(0)\rangle \neq 0$. The new phase γ_g is also projective-geometric and reduces to the Aharonov–Anandan phase ϕ_g in the case of cyclicity. In the case of an arbitrary open curve in $\mathscr{P}$, γ_g, when it is well-defined, is equal to ϕ_g defined on the geodesic closure of this curve – geodesics being defined of course relative to the mentioned connection on $\mathscr{P}$.

We may now ask whether γ_g is Galilean-invariant. In particular, suppose there is some gauge such that relative to the frame F the curve C in $\mathscr{P}$ defined by the Schrödinger evolution of the system in the interval $[0, T]$ is closed. A passive Galilean transformation to the frame F' implemented by eqn 19 will transform C into an open curve C'. Then is the geometric phase $\phi_g = \gamma_g$ for C equal to γ_g for C' (or equivalently ϕ_g for the geodesic closure of C')? The answer is no, and again the guilty party is the time-dependent exponent containing the canonical momentum $\mathbf{P}$ in eqn 19.[32]

[30] I thank Jeeva Anandan for this argument.
[31] See Anandan (1989).
[32] See Sjöqvist et al. (1997).

Intuitively, the non-invariance result is perhaps not surprising. A time-dependent translation of the coordinate axes (and hence a pure Galilean transformation) will, as we have seen, transform a closed curve in $\mathscr{P}$ into an open one, whose geodesic closure in turn is a different closed curve to the original one. After all, curves in $\mathscr{P}$ are geometric objects, but not in space-time. Given a rule for parallel transport in $\mathscr{P}$, why should we expect the anholonomy – the geometric phase – to be the same on both of these distinct closed curves? It is the curvature of $\mathscr{P}$ that is the relevant invariant entity, and this being the case we expect different anholonomies produced by parallel transport on different closed curves in the space.

Note that some commentators appear to regard the non-invariance result as representing a weakness in current formulations of geometric phase; indeed some attempts to remedy the situation have been undertaken.[33] However, the considerations expressed in the last paragraph, together with the fact that geometric phase has been experimentally 'observed' – even in cases not involving spin states which are Galilean invariant[34] – tend in my opinion to cast doubt on the necessity of an invariant reformulation of geometric phase. No doubt the issue deserves further analysis, but I leave it here.

I return now to the question of the ultimate significance of the discovery of the curved geometry of $\mathscr{P}$. (I accept of course that Berry's original 1984 discovery was genuinely surprising; it was generally assumed at the time that the total phase acquired in cyclic adiabatic evolution is purely dynamical.[35] The question I am now raising is: given the existence of geometric phases in this case and in the general case of Aharonov and Anandan, what is its significance?) The recognition that there is a feature of Schrödinger evolution that is indifferent to the dynamical details specified by the Hamiltonian – or at least the choice of Hamiltonian within an infinite relevant class – and that depends on only the fixed path in $\mathscr{P}$, has led to a significant geometrical reformulation of quantum mechanics. In particular, the symplectic structure of $\mathscr{P}$, and its role as a metric space have been clarified, leading *inter alia* to a new insight into the energy–time uncertainty relations.[36] The situation here

[33] A Galilean- (but not gauge-) invariant formulation of geometric phase was given in García de Polavieja (1997), and subsequently Bacciagaluppi (1997) has produced a formulation which is both Galilean- and gauge-invariant, defined on a bundle incorporating time and the ray space. An unsettling feature of the latter formulation is that the connection now depends on the scalar potential V in the Hamiltonian.

[34] For more details, see Sjöqvist et al. (1997).

[35] Even after Berry's discovery, it continued to be widely supposed that the geometrical phase associated with *non-cyclic* adiabatic motion could be ignored ('gauged away'), but this has been shown to be erroneous; see García de Polavieja & Sjöqvist (1998).

[36] See Anandan (1991).

is reminiscent of the discovery by Minkowski of the geometric formulation of special relativity, and the structure of Minkowski spacetime.[37] *But in both cases, no new predictions are involved.* (The interference effects that are involved in actual 'measurements' of geometric phase are of course predictable on the basis of good, old Schrödinger dynamics.) Indeed, the real significance of Minkowski's contributions to relativity theory only came to be seen in the later success of Einstein's general theory, where the spacetime geometry itself became a dynamical player. Analogously, the full significance of geometric phase may prove to be in its heuristic power in a future reformulation of quantum principles.[38]

4. Objectivity of sharp values?

A variety of interpretations of quantum mechanics seek to augment the standard state description of a quantum system (be the state pure or mixed) by specifying state-dependent rules for assigning sharp values to some of the self-adjoint operators representing magnitudes (or equivalently for assigning bivalent truth values to some propositions describing properties of the system). In what sense do these values (or truth-values) represent or reflect the existence of objective elements of reality? In this section, I shall briefly summarize some of the considerations contained in a recent attempt to answer this question.[39]

Let us consider those interpretations whose value-assignment rules coincide with the eigenstate–eigenvalue link (EEL) for systems which for some suitable period of time may be regarded as unentangled with the environment. (This includes some prominent versions of the so-called 'modal' interpretation.) To fix our ideas, let us imagine the free particle p in the original 1935 Einstein–Podolsky–Rosen argument, whose separated twin particle – originally entangled with p – has just undergone a measurement of the x-component of momentum P_x. (Recall that the initial EPR state of the two particles is an eigenstate of total momentum in a direction we associate with the x-axis.) Using either the EPR criterion of reality, or the stronger EEL (since we may effectively consider p to be in a pure (improper) eigenstate of momentum as a result of the distant measurement), P_x is assigned a sharp value for p. Classically, momentum is of course frame-dependent; classical

[37] I have compared the 'causal' properties, or rather the lack thereof, of both Minkowski spacetime and the projective Hilbert space in Brown (1996).
[38] See the concluding remarks in Anandan (1991, 1992).
[39] See Brown, Suárez & Bacciagaluppi (1998).

momentum is a relational property involving the body in question and a given inertial coordinate system (or a family thereof adapted to a given inertial frame) and under a passive Galilean transformation it changes its sharp value as in eqn 16 above. In quantum mechanics, the canonical operators **P** and **Q** can be taken to transform invariantly under a Galilean transformation (as was mentioned in section 2b above) and it transpires, unsurprisingly, that if P_x is sharp for p relative to the EPR lab frame, it is also sharp for p relative to a moving frame – the value having undergone the usual Galilean transformation for momentum. An exactly analogous situation holds for the position operator **Q**, as well as for the velocity operator $\dot{Q}_i \equiv i[H, Q_i]$, $i = x, y, z$, where H is again the Hamiltonian; indeed it is so even when the system is not free (and hence when the velocity operator is not proportional to the corresponding component of canonical momentum, failing thereby to transform invariantly).[40]

In all these cases, 'sharp values', when they obtain under EEL, are coordinate-dependent, and hence if construed as objective elements of reality, should presumably be regarded not as intrinsic properties of the quantum system but relational properties, analogously to their classical counterparts. (So far we are treating the properties as defined relative to inertial coordinate systems, but as is emphasized in the next section, this is arguably somewhat unrealistic from an operational point of view. For the moment, however, let us stick to this viewpoint.) Note however that when the particle is being acted upon by an external field that contributes a vector potential to the Hamiltonian, the canonical momentum is gauge-dependent. Classically, its value changes under a gauge transformation; but in quantum mechanics, the *very sharpness* of a given component of momentum is generally not preserved under a gauge transformation. Sharpness itself is not gauge-inde-

[40] Given the Hamiltonian in eqn 4 and eqn 17, the velocity operator transforms invariantly (i.e., $\dot{Q}'_i = \dot{Q}_i$) normally only at the initial instant when the two coordinate systems associated with a Galilean transformation coincide. It is worth noting here that a *derivation* of this generic Hamiltonian based on Galilean kinematics was provided by Jauch (1964), and has since repeated on a number of occasions in the literature. Jauch's theorem rests *inter alia* on the very assumption that $\dot{Q}'_i = \dot{Q}_i$ at $t = t' = 0$. Doubts about the *a priori* validity of this assumption, as well as about further details of the Jauch theorem, are found in Brown & Holland (1999). On a more general note, it can be argued that Jauch's approach, which uses the fundamental properties of space and time to constrain the form of the Hamiltonian, is pointing in the wrong direction. Spacetime structure itself may be seen as a consequence of the symmetries of the dynamics of quantum systems, and not as fundamental. This view is consistent with the profound analysis of the role of geometry in physical theories given in Anandan (1980); its implication in the case of quantum theory are further explored in Anandan (1997, particularly ch. 5).

pendent in quantum mechanics, a state of affairs that can also occur for the total energy when the gauge transformation is time-dependent.

This non-invariance of sharpness is exhibited in quantum mechanics also in the case of Galilean transformations, one example being that of orbital angular momentum (whose corresponding operator has a discrete spectrum and is therefore a more satisfactory object for the application of EEL than the magnitudes being discussed so far). A sharp component of angular momentum will lose its sharpness even when the spatial direction along which it is defined is orthogonal to the direction of the boost, and even at the instant when the systems of spatial axes associated with the two frames coincide. Note incidentally that it is not just boosts that produce this kind of situation; passive spatial translations of the coordinate system can cause some sharp, discrete observables (such as angular momentum again, or coarse-grained position) to become unsharp.

There seems, in short, to be even more reason in quantum mechanics than in classical mechanics generally to regard sharp values of properties as relational, as opposed to intrinsic attributes of the system. This is not the end of the story, but in the meantime it is worth emphasizing that the ways in which operators associated with distinct magnitudes transform under gauge or coordinate transformations can be quite different – ways which depend on the physical meaning of the magnitudes, and on their role within the quantum dynamics. The symmetry of the Hilbert space can look decidedly misleading when the implications of the symmetries of space and time are taken into account in quantum mechanics.

5. Quantum frame bodies and relational realism

In the previous section we were considering magnitudes, such as position, momentum and angular momentum, defined relative to inertial coordinate systems. In practice, measurements of such magnitudes do not occur (except occasionally in an approximate sense). Actual measurements establish relations between bodies.

Consider a rigid, impenetrable box located somewhere in space; in non-relativistic quantum mechanics it is a gauge- and coordinate-independent issue as to whether a quantum particle is wholly inside the box at any instant. The projection operator whose bivalent values correspond to the two possibilities (1 = 'wholly in', 0 = 'otherwise', say) must transform unitarily, so that if the 'in?' observable has a sharp value according to

EEL relative to one gauge or frame, it has the same sharp value under the relevant transformation.

All this seems straightforward, but note that the role of the box is that of a classical body. What if we treat it quantum mechanically? Indeed, what are the implications for our present discussion if we treat the entire laboratory (even when construed as moving inertially) as a quantum mechanical system?

Let us imagine then an entire closed laboratory, denoted by L, replete with rulers and clocks rigidly fastened to its walls, along with other equipment. Inside the lab an investigation is taking place of the behaviour of some microsystem p; we denote the system comprising the laboratory minus the particle by L_{-p}. Let us further suppose that with respect to some 'external' inertial reference frame F, the state of the laboratory L – which is assumed for simplicity to move freely – is at some instant of time an eigenstate of the x-component of total momentum relative to some Cartesian coordinate system adapted to F. (It might be imagined that this state is the result of an external measurement of the momentum of L completed at that instant.) Indeed, we might have it that L is at rest relative to F. Given the additive nature of linear momentum, it can easily be shown[41] that the (reduced) density matrix associated with the subsystem p at the instant in question describes a mixture of eigenstates of the operator (defined in the p factor Hilbert space) corresponding to the x-component of momentum. The particle is therefore wholly unlocalized relative to the x-axis of F. Hence it might seem that for an observer enclosed within the laboratory there is no possibility of 'seeing', or preparing, the particle as a localized wavepacket, concentrated in any specific region in the laboratory.[42]

But note that at the instant in question L is also wholly unlocalized, and it does not follow that *relative to the rigid walls of* L_{-p} the particle p cannot be considered strictly localized. Thus, the observer enclosed within L can resort to a variable representing the momentum of p relative to the much more

[41] See Lubkin (1970).

[42] This thought-experiment is of course highly idealized. No actual external measurement can determine the momentum of L with absolute accuracy, and even if the intervention were to lead to significant, macroscopic indeterminacy in the position of the centre of mass of L, decoherence brought about by interaction of the system with the environment (if only the cosmological background radiation) would occur with extreme rapidity, resulting in a mixture of sufficiently localized states of the lab relative to F. But we are dealing here with a question of principle, and we may restrict out attention to the precise instant at which supposed de-localization of L occurs. Furthermore, if desired we could imagine an external measurement of the position of the laboratory, in which case it is the ability to create plane waves (momentum eigenstates) for the subsystem p that would now be open to question.

massive system L_{-p} (which acts as a momentum 'reservoir' in the apt terminology of Lubkin);[43] such a variable is, unlike the total momentum of L relative to F, not strictly a conserved quantity. States associated with coherent superpositions of this 'relative momentum' can be effectively attributed to the particle p, which now represent localized wavepackets in the representation associated with the Fourier inverse of the relative momentum – the relative coordinate of the particle.

The upshot is that when the large system L_{-p} is treated as a quantum frame body, meaningful localization of the particle is possible even when L is sharp in momentum relative to the external frame F. Observe that the 'laboratory frame' defined by L_{-p} must be considered quite distinct from F, even when L is at rest relative to F. Indeed, we might think of F itself as being associated with, or modelled by some separate, even larger quantum system (a momentum reservoir of astronomical proportions?) so that the 'conserved' total momentum of L is now analogously considered a relative momentum defined with respect to this larger system.

The theory of 'quantum reference frames' has been developed further by Aharonov & Kauffher (1984), who posed a paradox (in one spatial dimension) based on our intuitions concerning the relativity principle. Ever since Galileo, we have come to expect that what transpires inside the inertial laboratory should not depend on its collective state of motion. In the context of quantum mechanics, an extension to this relativity (or 'equivalence') principle strongly suggests itself – one arguably implicit also in Lubkin's 1970 analysis: no observable processes occurring purely within L should demonstrate whether or not the state of L itself, relative to an external frame, is an eigenstate of centre-of-mass (c.o.m.) velocity or c.o.m. position. And yet the coordinate of the particle p defined relative to L_{-p} apparently fails to commute with the c.o.m. velocity of L relative to the external frame F, as does the velocity of the particle p relative to L_{-p} with the c.o.m position of L, when the mass of L_{-p} is finite.

The failure of these commutation relations rests, as Aharonov and Kauffher note, on the seemingly natural assumption that the relevant velocities of the free bodies p, L_{-p}, etc. are proportional to their respective canonical momenta. The authors further demonstrate that the difficulty is resolved if, for example, the particle p feels the presence of a vector potential which, in their words, represents the 'kick-back' of the finite mass reference

[43] See Lubkin (1970), where the analogous and more familiar problem of accounting for the localizability of the electron within a hydrogen atom is considered, when the atom as a whole is in a state of sharp momentum.

frame. The particle's velocity relative to L_{-p} is now proportional to its 'mechanical momentum' which of course depends on the mentioned vector potential as well as the canonical momentum. The vector potential is inversely proportional to the mass of L_{-p} and depends also on the momentum of the external frame F (itself also treated quantum mechanically) relative to L_{-p}. Although the vector potential in this case gives rise to no forces (as one would expect), when it is taken into account – as well as the analogous vector potential felt by F – the desired commutation relations are restored, and the 'paradox of the quantum reference frame' is resolved.

The 1984 study of quantum reference frames by Aharonov and Kauffher goes considerably further than is indicated here, and certainly deserves more critical attention by philosophers of physics than it has received to date.[44] But the study is consistent with the present theme that direct measurements of observables like position, velocity and momentum involve establishing relations between the object system p and some other material 'frame' *body* (such as L_{-p}), rather than an abstract coordinate system. This is not to say that inertial coordinate systems are not essential within the theoretical analysis; no accessible frame bodies with their attached physical clocks can strictly replace, or stand in for them, either in classical or quantum mechanics, except in an approximate sense.[45] Yet it is more realistic in relation to actual experiments to consider the observables under consideration as *operationally* defined in relation to such frame bodies. One lesson we may take from the Lubkin and Aharonov–Kauffher studies is that in standard quantum mechanics the particle p may in principle have a sharp location, for instance, relative to L_{-p}, but no sharp location relative to an inertial coordinate system or (more to the point operationally) to an external frame body, at the same instant.

I think this point serves to provide more operational grounds for the view that emerged in the last section. If, in the hope of providing an ontological interpretation of quantum mechanics, we introduce *state-dependent* rules for assigning sharp values to magnitudes associated with a specific quantum system, we should recognize that the objective status of such sharp values

[44] I am unaware of any analysis of this suggestive, but not wholly transparent study in the philosophical literature.

[45] Indeed, it is not clear to me that due recognition of the irreducible role of inertial coordinate systems in exactly defining the dynamics of quantum systems is given in Aharonov & Kauffher's 1984 analysis. (For further discussion of this role in the context of classical mechanics, see Barbour (1989, ch. 12) and Brown (1997, sect. 2).) It might for this reason be better to refer to the theory of quantum frame bodies, rather than quantum reference frames.

is relational, not absolute. The full implications of this state of affairs for the standard formulation of the measurement problem in quantum mechanics (in which the sharp positions of the generic 'pointer' of the apparatus are normally required to be observer-independent elements of reality) are, perhaps, still not widely appreciated.[46]

Appendix

Consider a spacetime manifold equipped with a flat affine connection, and a curve $x^i(\lambda)$ associated with the affine parameter λ. Relative to an inertial coordinate system, the geodesic equation for this curve (i.e. the equation of motion of a free particle) takes the form of Newton's first law:

$$\frac{d^2x^i}{d\lambda^2} = 0. \tag{A1}$$

In analogy with the first part of the minimal coupling procedure outlined in section 2.1, we wish to obtain the familiar generally covariant form of this equation,

$$\frac{d^2x^i}{d\lambda^2} + \Gamma^i_{kj}\frac{dx^k}{d\lambda}\frac{dx^j}{d\lambda} = 0. \tag{A2}$$

explicitly by way of replacing the ordinary partial derivative $\partial_j = \partial/\partial x^j$ by the covariant derivative associated with the connection coefficients Γ^i_{kj}, which may or may not be symmetric. Now let us write the vector field $dx^i/d\lambda$ defined on the curve in question as V^i; so eqn A1 becomes

$$\frac{dV^i}{d\lambda} = V^j\partial_j V^i = 0. \tag{A3}$$

Now replace the partial derivative ∂_j in eqn A3 by the covariant derivative D_j, which acts as $\mathrm{D}_j V^i = \partial_j V^i + \Gamma^i_{kj}V^k$. We then obtain

$$V^j(\partial_j V^i + \Gamma^i_{kj}V^k) = 0. \tag{A4}$$

which is equivalent to eqn A2.

[46] It may well be that the theory of quantum frame bodies is largely compatible with the interpretation of quantum mechanics defended recently in Mermin (1998), although I do not follow Mermin in his dismissal of both the measurement problem and the many-worlds interpretation on the basis of the wholly ineffable nature of consciousness from point of view of physics (Mermin 1998, sect. VIII).

References

Aharonov, Y. & J. Anandan (1987). 'Phase Change During a Cyclic Quantum Evolution', *Physical Review Letters* **58**, 1593–6.

Aharonov, Y. & T. Kauffher (1984). 'Quantum Frames of Reference', *Physical Review D* **30**, 368–85.

Aitchison, I. J. R. & K. Wanelik (1992). 'On the Real and Complex Geometrical Phases', *Proceedings of the Royal Society of London A* **439**, 25–34.

Anandan, J. (1980). 'On the Hypotheses Underlying Physical Geometry', *Foundations of Physics* **10**, 601–29.

(1989). 'Geometry of Cyclic Evolutions', in M. Kafatos (ed.), *Bell's Theorem, Quantum Theory and Conceptions of the Universe*, Dordrecht: Kluwer.

(1991). 'A Geometric Approach to Quantum Mechanics', *Foundations of Physics* **21**, 1265–84.

(1992). 'The Geometric Phase', *Nature* **360**, 307–13.

(1997). 'Reality and Geometry in Quantum Theory', DPhil. thesis in philosophy, Oxford University.

Anandan, J. & H. R. Brown (1995). 'On the Reality of Space-Time Geometry and the Wavefunction', *Foundations of Physics* **25**, 349–60.

Bacciagaluppi, G. (1997). 'Making geometric phase both Galilean- and Gauge-Invariant', manuscript.

Ballentine, L. E. (1990). *Quantum Mechanics* (New Jersey: Prentice Hall).

Barbour, J. (1989). *Absolute or Relative Motion?* Vol I: *The Discovery of Dynamics* (Cambridge: Cambridge University Press).

Berry, M. V. (1984). 'Quantal Phase Factors Accompanying Adiabatic Changes', *Proceedings of the Royal Society of London A* **392**, 45–57.

Berry, M. V. and J. M. Robbins (1997). 'Indistinguishability for Quantum Particles: Spin, Statistics and the Geometric Phase', *Proceedings of the Royal Society of London A* **453**, 1771–90.

Brown, H. R. (1996). 'Bovine Metaphysics: Remarks on the Significance of the Gravitational Phase Effect in Quantum Mechanics', in R. Clifton (ed.), *Perspectives on Quantum Reality* (Dordrecht: Kluwer), pp. 183–93.

(1997). 'On the Role of Special Relativity in General Relativity', *International Studies in the Philosophy of Science* **11**, 67–81.

Brown, H. R., A. Elby & R. Weingard (1996). 'Cause and Effect in the Pilot-Wave Interpretation of Quantum Mechanics', in J. T. Cushing et al. (eds), *Bohmian Mechanics and Quantum Theory: An Appraisal*, (Dordrecht: Kluwer), pp. 309–19.

Brown, H.R. & P. R. Holland (1998). 'The Galilean Covariance of Quantum Mechanics in the Case of External Fields', *American Journal of Physics*. **67**, 204–14.

Brown H. R., E. Sjöqvist & G. Bacciagaluppi (1999). 'Remarks on Identical Particles in de Broglie-Bohm theory', *Physics Letters A*. **251**, 229–35.

Brown H. R., M. Suárez & G. Bacciagaluppi (1998). 'Are "Sharp Values" of Observables Always Objective Elements of Reality?', in D. Dieks & P. Vermaas (eds.), *The Modal Interpretation of Quantum Mechanics*, (Dordrecht: Kluwer) pp. 289–306.

Brown, H. R. & R. Sypel (1995). 'On the Meaning of the Relativity Principle and Other Symmetries', *International Studies in the Philosophy of Science* **43**, 381–407.

Collins, P. D. B., A. D. Martin & E. J. Squires (1989). *Particle Physics and Cosmology* (New York: John Wiley & Sons).

Fonda, L. & G. L. Ghirardi (1970). *Symmetry Principles in Quantum Physics* (New York: Marcel Dekker).

García de Polavieja, G. (1997). 'Galilean Invariant Structure of Geometric Phase', *Physics Letters A* **232**, 1–3.

García de Polavieja, G. & E. Sjöqvist (1998). 'Extending the Quantal Adiabatic Theorem: Geometry of Noncyclic Motion', *American Journal of Physics* **66**, 431–8.

Holland, P. R. (1993). *The Quantum Theory of Motion* (Cambridge: Cambridge University Press).

Jauch, J. M. (1964). 'Gauge Invariance as a Consequence of Galilei-Invariance for Elementary Particles', *Helvetica Physica Acta* **37**, 284–92.

Kuchař, K. (1980). 'Gravitation, Geometry, and Nonrelativistic Quantum Theory', *Physical Review D* **22**, 1285–99.

Lawrie, I. D. (1990). *A Unified Grand Tour of Theoretical Physics* (Bristol: Adam Hilger, IOP Publishing Ltd).

Le Bellac, M. & J.-M. Lévy-Leblond (1973). 'Galilean Electromagnetism', *Il Nuovo Cimento* **14B**, 217–33.

Leinaas, J. M. & J. Myrheim (1977). 'On the Theory of Identical Particles.', *Nuovo Cimento B* **37**, 1–23.

Lévy-Leblond, J.-M. (1976). 'Quantum Fact and Fiction: Clarifying Landé's Pseudo-Paradox', *American Journal of Physics* **44**, 1130–2.

Lubkin, E. (1970). 'On Violation of Superselection Rules', *Annals of Physics* **56**, 69–80.

Mermin, N. D. (1998). 'What is Quantum Mechanics Trying to Tell Us?', *American Journal of Physics* **66**, 753–67.

Mukunda, N. & R. Simon (1993). 'Quantum Kinematic Approach to the Geometric Phase. I. General Formalism', *Annals of Physics* **228**, 205–68.

Nakahara, M. (1990). *Geometry, Topology and Physics* (Bristol: Adam Hilger, IOP Publishing Ltd).

Rai Dastidar, T. K. & K. Rai Dastidar (1994). 'Gauge Invariance in Non-Relativistic Quantum Mechanics', *Il Nuovo Cimento* **109B**, 1115–18.

(1995). 'Local Gauge Invariance of Relativistic Quantum Mechanics and Classical Relativistic Fields', *Modern Physics Letters A* **10**, 1843–6.

Ryder, L. H. (1987). *Quantum Field Theory* (Cambridge: Cambridge University Press, paperback edition).

Sakurai, J. J. (1964). *Invariance Principles and Elementary Particles* (Princeton: Princeton University Press).

Sjöqvist, E., H. R. Brown & H. Carlsen (1997). 'Galilean Noninvariance of Geometric Phase', *Physics Letters A* **229**, 273–8.

Takagi, S. (1990). 'Equivalence of a Harmonic Oscillator to a Free Particle', *Progress of Theoretical Physics* **84**, 1019–24.

(1991). 'Quantum Dynamics and Non-Inertial Frames of Reference, I, II and III', *Progress of Theoretical Physics* **85**, 463–79; 723–42; **86**, 783–98.

Teller, P. (1997). 'Essay Review. A Metaphysics for Contemporary Field Theories', *Studies in History and Philosophy of Modern Physics* **28**, 507–22.

Valentini, A. (1997). 'On Galilean and Lorentz Invariance in Pilot-Wave Dynamics', *Physics Letters A* **228**, 215–22.

Wald, R. M. (1984). *General Relativity* (Chicago: University of Chicago Press).

4

The 'beables' of relativistic pilot-wave theory

SIMON SAUNDERS

As I write it is ten years since Michael Redhead's *Incompleteness, Non-Locality, and Realism* went to press. For philosophers of physics, it remains the one book that is on everyone's shelf. Achieving as it did a certain non-locality, or ubiquity, it was also incomplete: the most important omission was the de Broglie–Bohm or pilot-wave theory, which went unmentioned. It is my pleasure to say something about it here, specifically on the question of 'beables' in the relativistic case.

A widely held view is that the pilot-wave approach can make do with classical fields as beables. I argue that this is mistaken, and that a particle ontology is essential if the theory is to solve the problem of measurement. There are two familiar strategies, due to Feynman and Dirac; although each has its problems, Dirac's ideas are the more closely wedded to non-relativistic theory, so the more amenable to pilot-wave methods. The beables are particles, and, with opposite charge, the absence of particles, in an infinite plenum filling the void.

1. Non-relativistic quantum mechanics (NRQM)

It will be helpful to have the example of the non-relativistic theory spelt out in some detail. The beables are structureless point-particles. In the simplest case, where we have a single spinless particle, the Schrödinger equation referred to a Galilean frame is:

I am grateful to Jeremy Butterfield, Guido Bacciagaluppi and Anthony Valentini for stimulus and fruitful discussions. I would also like to thank an anonymous referee for several helpful suggestions. Finally, I would like to express my gratitude and indebtedness to Michael Redhead, for many years of guidance.

$$\left(-\frac{\hbar^2}{2m}\nabla^2 + V(\mathbf{x}, t)\right)\Psi(\mathbf{x}, t) = \frac{i\hbar\partial}{\partial t}\Psi(\mathbf{x}, t) \tag{1}$$

Familiar manipulations yield, for the real and imaginary parts of $\Psi = R\exp(iS)$:

$$\frac{\partial S}{\partial t} + \frac{(\nabla S)^2}{2m} - \frac{\hbar^2}{2m}\frac{(\nabla^2 R)}{R} + V = 0 \tag{2}$$

$$\frac{\partial R^2}{\partial t} + \nabla \cdot \left(\frac{R^2 \nabla S}{m}\right) = 0. \tag{3}$$

By Gauss's theorem, it follows from eqn 3 that the integral of $R^2(\mathbf{x}, t)$ over all space is constant in time (assuming R and ∇S vanish at infinity). Eqn 3 should be compared with the equation of continuity in classical hydrodynamics, with $\rho(\mathbf{x}, t)$ as the density function of a classical fluid, and $\mathbf{v}(\mathbf{x}, t)$ the velocity distribution function:

$$\frac{\partial \rho}{\partial t} + \nabla \cdot (\rho \mathbf{v}) = 0. \tag{4}$$

The spatial integral of the density function over a volume V is the total mass contained in V; eqn 4 states that any change in this mass must be compensated by a net flow of mass into V across its boundary ∂V, as given by the surface integral of the momentum density $\rho\mathbf{v}$ over ∂V. Evidently eqns 3 and 4 will have a similar interpretation if R^2 is taken to be the analog of ρ, the mass-density, and if the gradient of the action S is interpreted as a momentum density (as it is in classical Hamilton–Jacobi theory). We define a velocity field on the basis of this analogy:

$$\mathbf{v}(\mathbf{x}, t) = (1/m)\nabla S(\mathbf{x}, t). \tag{5}$$

What is its physical meaning? It is at this point that the pilot-wave theory and conventional quantum mechanics part company. According to the pilot-wave theory, there exist *particles with definite trajectories*. The *allowed* trajectories are the integral curves of the vector-field $\mathbf{v}$. So long as S is well-defined, these trajectories are continuous functions of the time, one and only one of which passes through any point at each time. Formally, we are to find a family of functions $\mathbf{y}(t)$ such that

$$\left.\frac{d\mathbf{y}}{dt}\right|_{t=t_0} = \mathbf{v}(\mathbf{x}, t_0)\Big|_{\mathbf{x}=\mathbf{y}(t_0)}. \tag{6}$$

Solving eqn 6 for given S at space-time point p with coordinates $(\mathbf{y}_0, t_0)$, so long as $\mathbf{v}$ at p and time t_0 is non-zero, we obtain a unique trajectory $\mathbf{y}_p$ through p with $\mathbf{y}_p(t_0) = \mathbf{y}_0$. The position of a particle determines its velocity, and hence its change in position $\delta\mathbf{y}_0 = \mathbf{v}(\mathbf{y}_0, t_0)\delta t$ in time δt after $t = t_0$; and thereby its entire trajectory.

If we suppose that only *one* trajectory actually exists (the one going through the point p, say), and that positions along this trajectory are what are observed in subsequent measurements, then obviously there is no measurement problem of the conventional kind. Regardless of whether or not the *state* evolves into a superposition of 'position eigenstates', there will be only *one* particle position at each time t, as fixed by $\mathbf{y}_p(t)$.

The method is easily generalized to n-particle systems, replacing throughout the coordinates $\mathbf{x} = (x_1, x_2, x_3)$ on the 1-particle configuration space by $\{x_i^r\}$, $r = 1, \ldots, n$, $i = 1, 2, 3$ on the $3n$-dimensional configuration space Γ (assuming the systems are unconstrained). Given that at time t_0 the system occupies the configuration space point $\gamma \in \Gamma$, we once again obtain a unique trajectory through γ (a map $\mathbb{R} \rightarrow \Gamma$ with $t_0 \rightarrow \gamma$), and with that unique positions at each time for each of the n particles.

In this way the measurement apparatus, too, can be modelled in the theory. It now *follows*, with no further assumptions, that the spacetime properties of macroscopic objects, including pointer positions, will be just as definite, no more and no less, as those of their constituent particles. At the very worst the pointer might be vaporized or otherwise broken into parts, but even in that case, the result will be describable in classical terms (because the behavior of the constituent particles will be classically describable). There can be no Schrödinger cat paradox in consequence.

It is another matter as to whether the statistics of outcomes match the values predicted, using the usual measurement postulates. To return to the 1-particle case, they will if the sample population is selected in accordance with the probability measure:

$$\rho(\mathbf{x}, t) = |\Psi(\mathbf{x}, t)|^2 = R^2. \tag{7}$$

If particle positions are 'typical', at time t_0, i.e. they have the distribution (7) at $t = t_0$, then they will be 'typical' at all other times (by virtue of the equation of continuity, eqn 3). What of the remaining equation, eqn 2? Solutions to this determine ∇S as a function on $\Gamma \times \mathbb{R}$, and thereby determine the trajectories $\mathbf{y}_p$ for variable p. We can, however, combine eqns 2 and 6 into a single second order equation. Differentiating eqn 6 with respect to time we obtain:

$$d^2\mathbf{y}/dt^2 = d\mathbf{v}(\mathbf{x}, t)/|_{\mathbf{x}=\mathbf{y}(t)} = \partial\mathbf{v}(\mathbf{x}, t)\partial t|_{\mathbf{x}=\mathbf{y}(t)} + (\mathbf{v}\cdot\nabla)\mathbf{v}(\mathbf{x}, t)|_{\mathbf{x}=\mathbf{y}(t)} \quad (8)$$

where the second term takes account of the variation in $\mathbf{v}$ in time δt due to the change in the value of $\mathbf{x}$, from $\mathbf{y}(t)$ to $\mathbf{y}(t+\delta t)$. Taking the divergence of eqn 2 we obtain:

$$(\partial/\partial t + (1/m)\nabla S\cdot\nabla)\nabla S = -\nabla(V+Q)$$

where Q, the 'quantum potential', is the term involving R in eqn 2. Using eqn 8 we obtain the equation of motion, eqn 6, as a second-order equation:

$$\frac{d^2}{dt^2}(m\mathbf{y})|_{t=t_0} = -\nabla((V(\mathbf{x}, t_0) + Q(\mathbf{x}, t_0))|_{\mathbf{x}=\mathbf{y}(t_0)}. \quad (9)$$

This has a form familiar from classical mechanics, but we must remember that the three components of the velocity at time t_0 are not constants of integration along with the position $\mathbf{y}_0$ at this time. Only those solutions $\mathbf{y}$ to eqn 9 for which eqn 6 holds at $t=t_0$ are allowed, given which eqn 6 then holds at all times; so the velocities are completely fixed. In NRQM, eqn 6 is the fundamental equation. Nothing more is needed, for the pilot-wave theory, other than the Schrödinger equation, and the use of the measure eqn 7. In particular, measurements are automatically taken care of. There is no need for any further postulates, or mention of experiment or observation. The measurement problem is solved.

2. Relativistic particles

Can we proceed analogously in the relativistic case? But we know that in orthodox quantum theory there is no *uniform* generalization of NRQM to the relativistic domain. Only in special circumstances can we carry over the usual prescriptions. One sees this very simply in the case of the scalar (Klein–Gordon) wave equation, with external potential A_μ, $\mu = 1, 2, 3, 4$ (putting $\hbar = c = 1$):

$$g^{\mu\nu}(i\partial_\mu - eA_\mu)(i\partial_\nu - eA_\nu)\psi = m^2\psi. \quad (10)$$

From this we deduce that there exists a divergence-free 4-vector j_s (writing $\varphi\overleftrightarrow{\partial}_\nu\psi$ for $\varphi\partial_\nu\psi - (\partial_\nu\varphi)\psi$) with components:

$$j_s^\mu = g^{\mu\nu}(i/2m)\psi^*(i\overleftrightarrow{\partial}_\nu - 2eA_\nu)\psi \quad (11)$$

but in the general case the vector j_s ('s' for 'scalar') is not everywhere time-like. Furthermore, whilst the spatial integral of its time-component is conserved, by Gauss's theorem, this integral need not be positive. So we cannot use eqn 11 to define an invariant positive-define inner-product on the space of solutions to eqn 10, blocking the construction of a Hilbert space and a probability interpretation. In the free case, when the external potentials are zero, one can restrict the class of solutions to define a Hilbert space with eqn 11 positive-definite (the Hilbert space of positive-frequency states). The same can be done for certain classes of slowly varying external potentials. In both cases the phenomenology is similar to that of NRQM; the creation and annihilation of particles, the processes characteristic of particle physics, involve strong fields and rapidly-changing potentials.

There are worse difficulties for the analogous 1-particle model in pilot-wave theory. The lack of a positive definite norm likewise means that there will be problems in the probability interpretation (cf. eqn 7), but instead of the problem of defining the Hilbert space, and the operator formalism that goes with it, there is a difficulty in defining the particle trajectories: the spatial part of eqn 10 ought to give us the guidance condition, but since j_s^0 is not everywhere timelike, there are times when some of the integral curves of j_s become spacelike. Unless these can be excluded, or shown to have measure zero, then particles can move at superluminal speeds, and even reverse their direction in time. This is so even in the free case, with $A_\mu = 0$, and even restricting ourselves to superpositions of states which have positive energy (in the usual sense), and even if these individually yield timelike j_s (Kyprianidis 1985).

Evidently the scalar 1-particle pilot-wave theory is in trouble, even in the kinematic limit, where the conventional 1-particle theory is free of any difficulty. This is a surprise; we expected that this naive approach would encounter difficulties in parallel to those of the standard formalism, but here we find them in the limit of free particles. Since this is the asymptotic limit of scattering theory in RQFT, routinely used in applications, the result is not encouraging. This conclusion is widely shared (Bohm & Hiley, 1993; Holland 1993): in the scalar case, no credible 1-particle pilot-wave theory is on offer.

What of the 1-particle fermion theory? Here too the conventional theory is in difficulties, but again there is a well-defined kinematic limit. Let us see how things stand with its pilot-wave analogue. We shall work with the Dirac equation. The scalar wave-function of NRQM is replaced by a 4-component function $\Psi : x \to \mathbb{C}^4$, $x \in \mathbb{R}^4$, transforming under a finite-dimensional (non-unitary) representation of the group SL(2,C) (the covering group of the Lorentz group), with 4×4 matrix generators γ^μ, $\mu = 1, 2, 3, 4$, obeying the relations:

$$\gamma^\mu \gamma^\nu + \gamma^\nu \gamma^\mu = 2g^{\mu\nu}.$$

In terms of the γ-matrices the Dirac equation for external potentials A_μ is:

$$\gamma^\mu(i\partial_\mu - eA_\mu)\Psi = m\Psi. \tag{12}$$

The adjoint representation of SL(2,C) is in terms of bispinors $\tilde{\Psi}$; referred to a Lorentz frame (where "†" denotes transpose and complex conjugation):

$$\tilde{\Psi}(\mathbf{x}, t) = (\gamma^0\Psi)^\dagger(\mathbf{x}, t). \tag{13}$$

Using these representations, we can construct the divergence-free 4-vector j_{D} (summing over spinor indices) with components:

$$j_{\mathrm{D}}^\mu(x) = \tilde{\Psi}(x)\gamma^\mu\Psi(x). \tag{14}$$

This gives us a real 4-dimensional vector field which, in contrast to eqn 10, is everywhere timelike. We can therefore define a conserved positive-definite inner-product:

$$< \Psi, \phi >= \int \tilde{\Psi}(x)\gamma^\mu\Phi(x)d\sigma_\mu. \tag{15}$$

This does not really solve the problems encountered in the scalar case, however; here, just as in the scalar case, the naive Hamiltonian, as determined by the wave-equation, has negative eigenvalues. The situation is not as bad as in the KG theory, where we cannot define the Hamiltonian as an operator on a Hilbert space at all, but it is bad enough: if there is no lower bound to the energy, there can be no ground state stable under strong couplings; at best there will only be a local minimum, and the most that could be hoped for is a perturbative treatment with respect to it.

What about the pilot-wave theory? It is clear that the previous difficulty, that afflicted the scalar guidance condition, is no longer a problem. The Dirac current j_{D} is everywhere timelike, unlike the scalar current j_{S}. Its integral curves are at least candidates for particle trajectories. First form the 4-velocity field with components:

$$u^\nu(x) = f(x)j^\nu(x)$$

(where $f(x)$ is a normalization factor ensuring that $u^\nu u_\nu = 1$ for non-zero j). Its spatial part $\mathbf{v}$, a 3-velocity field, has components $v^k = u^k/u^0$, $k = 1, 2, 3$:

$$v^k(x) = \frac{\tilde{\Psi}(x)\gamma^k\Psi(x)}{\Psi^\dagger(x)\Psi(x)}. \tag{16}$$

The particle trajectories now follow from this as in NRQM, using eqn. 6. It is easy to check that $|\mathbf{v}|^2$ is bounded by 1 (i.e. c, the velocity of light), as it should be. If, further, we use eqn 15 to define a probability interpretation, we see that the ensemble average of the components of velocity are:

$$< v^k >= \int \Psi^\dagger(x)\alpha^k\Psi(x)\mathrm{d}^3x \qquad (17)$$

where $\alpha^k = \gamma^0\gamma^k$, from which we see that the expectation value of the speed is likewise bounded by c. The important point is that all of these favourable features hold whether or not the state is made up of positive or negative frequency parts.

It seems, in fact, that the pilot-wave theory has done much better than the standard 1-particle theory. For example, whilst eqn 17 has often been interpreted as the mean electron velocity, in the orthodox tradition, there has always been the problem that the matrices α^k do not commute; if these are the velocity operators, it would follow (by the usual arguments) that different components of the velocity are not simultaneously measurable. Their eigenvalues, moreover, are ± 1, giving rise to Schrödinger's picture of 'zitterbewegung', of particles trembling back and forth at the speed of light.

These have remained puzzles within conventional thinking. One solution, adopted by many, is to give up on the concept of particle positions and velocities, and make do with momenta instead, which are perfectly simple and easy to define by self-adjoint operators (see my 1991 for references and further discussion). But they are not puzzles for the pilot-wave theory, which defines the positions and velocities by the guidance condition, independent of the Hilbert-space theory altogether. If there are no covariant position operators, then so much the worse for *operators*; the pilot-wave theory can make do without them.

But success in this context brings with it a new difficulty. The 1-particle pilot-wave theory had better not be free of *all* internal difficulties, for if it is then it will be applicable whatever the external potentials. But we know that when the field strengths are large or quickly varying then the particle number is likely to change. We know this from the conventional theory, but also from experiment. The 1-particle pilot-wave theory better *had* get into difficulties, given sufficiently strong couplings. If we look again to the history of the subject, the problem that Dirac and others were wrestling with was only indirectly related to the existence of position operators (which remains a problem to this day). Rather, he was concerned with the interpretation of the negative-frequency states; just what the pilot-wave theory appears to be blind to. Only on the standard interpretation does this notion of negative total energy appear so peculiar, and loom so large: one expects the potential energy to vanish in the asymptotic limit, but then if the total energy were negative, the kinetic energy would have to be negative. We have no idea what that would mean. The theory cried out to be replaced by one better, and it was clear that the distinction between positive and negative frequency solutions was the key.

What seemed to be a strength of the pilot-wave theory now looks to be its failing; in terms of the beables of pilot-wave theory, there can never be any question of negative kinetic energies; particle trajectories are always well-defined, independent of whether the state has negative frequency parts. The difficulty is more concealed, for it lies in the quantum potential. Appear it must, if we have the same unitary evolution of the state in the two theories, for things *do* go wrong at the level of the state, on conventional thinking. On the standard view, the absence of a lower bound to the energy will be catastrophic if the couplings are sufficiently strong, for components of state with arbitrarily large negative energies will grow in amplitude at the expense of all others. That ought to make for unbounded growth in the (positive) kinetic energies recognized by the pilot-wave theory, those defined by eqn 16. But it is noteworthy that this reasoning depends on the energy as defined by the state, and hence the operator formalism; if we are to reproduce it in the pilot-wave theory, it will be in terms of the quantum potential.

Evidently the successes and the failings of the pilot-wave theory, in the 1-particle interpretation, are different in kind from those of the conventional approach. It is far from clear how the concept of anti-matter, wedded as it is to negative energies in the standard theory, is to make its appearance at the level of the beables. Of course, in view of the ubiquity of particle creation and annihilation processes, as actually observed, we already know that the dynamics should lead to a change in particle number, a concept foreign to NRQM and, as we have seen, to the pilot-wave theory.

3. Alternative particle beables

Historically speaking, regarding the negative-energy difficulty, three strategies turned out to be productive. The first was Dirac's hole theory, of the early 1930s. This led him to predict the existence of antimatter, specifically the positron. The second strategy, also built on Dirac's work, was to reformulate the theory as a theory of fields (RQFT), as was more common by the late 1930s. The third proposal was due to Feynman, which again emphasized the particle aspect. Since for the moment we are considering particle interpretations, we are concerned with the first and third strategy.

The basic idea of Dirac's hole theory applies only to fermions, particles subject to the Pauli exclusion principle. Dirac suggested that not only do negative-energy particles exist, but that they exist in such abundance that all the negative energy states are occupied. This is the Dirac negative energy sea. Since, according to the exclusion principle, no two fermions can occupy the same state, transitions to negative energy states would in general be prohibited. The only exception would be if a negative energy electron were to acquire sufficient energy so as to be ejected from the sea, acquiring positive total energy. Since the total energy includes the rest-mass energy, there is always a considerable threshold energy required for this to happen. But if it does happen, then there will be a 'hole' in the negative-energy sea, and it will be possible for a positive energy particle to make a transition into the hole.

Dirac showed that such a hole would behave as though it were a positive-energy particle of positive charge. So the creation of a hole would appear to be the creation of *two* positive-energy particles, of opposite electric charge; and the filling of a hole would correspond to the *disappearance* of the hole and the particle, so the annihilation of oppositely charged particles. In this way Dirac was led to postulate the existence of annihilation and creation processes, involving oppositely charged pairs. Since the mechanism obviously applies to any sort of particles, whether or not they are electrically charged (so long as they are fermions), we should really keep track of the number of holes and positive energy particles using a more universal notion of charge, positive for *particles*, negative for *antiparticles*.

The hole theory makes it obvious that high-energy phenomena will involve variable particle number, despite the fact that the total number of positive and negative energy particles, all infinitely many of them, will be conserved, in accordance with the structure of NRQM. From a formal point of view the theory is conservative. That would seem to auger well for the pilot-wave theory, which works so well in NRQM. And Dirac's ideas can certainly get a foothold: one might hope to define change in particle number accordingly, at least in a perturbation sense, distinguishing positive and negative frequency parts by reference to the free Hamiltonian (with zero external potential). Ordinary QED can do no better, after all. But how the exclusion principle, or something analogous to it, would be related to a condition on the beables, is not so clear. This is at best a programme for research.

Feynman's strategy was very different. He supposed that particles, or more properly probability amplitudes for particle processes, could be associated with paths that zig-zagged backwards and forwards in time. Such reversals in the path could be reinterpreted just as were Dirac's transitions into and out of the negative energy sea. A particle going backwards in time would appear to be an oppositely-charged particle moving *forward* in time;

and on reversal back to the forward direction, it would appear that two particles had been created, one with opposite charge to the other. Conversely, a particle moving forward in time, reversing into the backwards direction, would appear as an instance of pair annihilation. Like the Dirac hole theory, it is clear that the apparent particle number can only change by multiples of two, that the mechanism applies quite generally, and that the *real* number of particles present (the number zig-zagging backwards and forwards in time) may be very different from the number observed. As Feynman said, it may be that there is only a *single* electron.

If Dirac's theory has too many things in it, Feynman's has too few. But the important question is whether the pilot-wave theory can make use of the idea. Of course Feynman's particles, though associated with trajectories (through the path-integral formalism), were not 'beables' in any classical sense, no more so than were Dirac's. Neither of them offered a realistic theory which could clear up the problem of measurement. But his picture could surely apply to the trajectories of the pilot-wave theory, if only they have the appropriate form.

Alas, they clearly don't, if they are the integral curves of the Dirac current j_D; what was a strength becomes its weakness. Should we work with the scalar KG current j_s instead? But we need a theory of pair creation and annihilation above all for fermions, and not for scalar particles. Alternative 4-vector fields have been proposed in the literature (Bohm, Schiller & Tiomno, 1955), but they do no better in this regard, and worse in others (Bohm & Hiley, 1993, sect. 10.3). The root problem is that Feynman's methods, based as they are on the path-integral approach to quantization, have a more tenuous relation to the Schrödinger equation, the point of departure of the pilot-wave theory, and give us no clues as to how they can be understood in canonical terms.

In point of fact, Dirac's methods were not as fruitful as those of field theory, whilst Feynman's, in the more fundamental applications, likewise appealed to Lagrangians defined by locally interacting fields. Finally, we come on to field theory.

4. Fields as beables

In quantum *mechanics*, the beables are *particles*; in quantum *field* theory, they should be *fields*. This is surely the most natural strategy. To make contact with the familiar methods of the standard formalism of field theory, we shall begin with the pilot-wave theory of normal modes. We take the field configuration space to be parameterized by a denumerable infinity of coordinates (q_k), and rework the basic steps of the pilot-wave formalism using the state as a function on this space (that is, it is a 'functional', since a complex function of infinitely many coordinates is essentially a map from a Hilbertian function space to the complex numbers). The normal mode coordinates are essentially the real part of the complex amplitudes of the normal modes. Using box normalization in volume V:

$$\psi(\mathbf{x}) = \frac{1}{V^{1/2}} \sum_k a_k \exp(i\mathbf{k}.\mathbf{x}) \tag{18}$$

(here $\mathbf{k}.\mathbf{x} = \Sigma_i k_i x_i$, and $k_i = 2\pi n_i/L$, $n_i \in \mathbb{Z}^+$, $i = 1, 2, 3$, for a box of side L); we use the real c-numbers

$$q_k = |\mathbf{k}|^{-1}(a_k + a_k^*) \tag{19}$$

as coordinates for the field configuration space. The phase space can be coordinatized by these and their canonical momenta. In the pilot-wave formalism, we write the q_k's as functions of the time, with time-derivatives given as before as tangents to the vector field (in configuration space) defined by the phase S, now a functional of the infinitely many parameters q_k (equivalently, of ψ). Formally we can follow through the steps (5), (6) to obtain:

$$\left.\frac{dq}{dt}k^{(t)}\right|_{t=t_0} = \left.\frac{1}{m}\frac{\partial S}{\partial q_k}[q_1, \ldots, q_k, \ldots]\right|_{q_k=q_k(t_0)} . \tag{20}$$

S is as before the phase of the state Φ, the solution to the Schrödinger equation (not to be confused with the field equation for ψ):

$$H\Phi[\psi, t] = i\hbar\frac{\partial}{\partial t}\Phi[\psi, t] \tag{21}$$

where H is the Hamiltonian (involving the usual field operators), and square brackets indicate that the state Φ is a functional of the field ψ.

Eqn 20 specifies the kth component of an (infinite-dimensional) vector field over an (infinite dimensional) space as a function of the time. Viewing S as a functional $S[\psi, t]$ over the space of classical field configurations ψ, this function of the time is the change in S in the 'direction' $\exp(i\mathbf{k}.\mathbf{x})$ at time t.

This approach has been applied to some simple kinematical problems in pure electromagnetic field theory. As it stands we can see the problems we will encounter in trying to extend it to the interacting case: we will no more be able to solve eqn 21 and express S as a functional in normal modes, in the interacting case, than we can make use of the plane-wave expansion, in the ordinary approach. But of course in the first place we should aim for a perturbative treatment, as in the usual theory.

Here we are trying to get an overview of the general structure of the theory. To this end it will be helpful to reformulate it in more abstract terms, without relying on normal modes. We may use the field values themselves, parameterized by coordinates on $\mathbb{R}^3$. Formally we can view this as the replacement of an orthonormal basis (the normal modes) by delta functions. If we do this, we obtain the time derivative of the field ψ at a point $\mathbf{x}$ of $\mathbb{R}^3$ (rather than the kth component of the field) in terms of the functional derivative of the phase in the 'direction' $\delta_{\mathbf{x}}$ (where $\delta_{\mathbf{x}}(\mathbf{y}) = \delta(\mathbf{x}-\mathbf{y})$).

Let us make this precise. The functional derivative of a functional $A[\ ]$ at χ in the direction ψ is:

$$\frac{DA}{D\psi}[\chi] = \lim_{\varepsilon \to 0} \frac{A[\chi + \varepsilon\psi] - A[\chi]}{\varepsilon}.$$

In the case of the direction $\delta_{\mathbf{x}}$ we use the notation:

$$\frac{\delta}{\delta\chi(\mathbf{x})} A[\chi] \underset{\text{def}}{=} \frac{DA}{D\delta_{\mathbf{x}}}[\chi].$$

Eqn 20 is then replaced by:

$$\left.\frac{\partial\psi(\mathbf{x}, t)}{\partial t}\right|_{t=t_0} = \left.\frac{\delta}{\delta\chi(\mathbf{x})} S[\chi, t_0]\right|_{\chi(.)=\psi(.,t_0)}. \tag{22}$$

This is the most natural generalization of the ideas sketched at eqns 5, 6, for the non-relativistic 1-particle case, to field theory.

In principle, if we can solve eqns 21, 22, we would have for each choice of beable at time $t = t_0$ – for each field configuration ψ at time t_0, as a complex-valued function on $\mathbb{R}^3$ – a trajectory through the space of classical field configurations. We would have the beable at every other time. In principle this problem is mathematically well-posed in the interacting case; in ϕ^4 theory, for example, in 1 or 2 spatial dimensions.

What of fermion fields? The standard field configuration space in this case is the space of bispinor functions whose values are not complex numbers but complex Grassmann numbers (see Berezin, 1966). These are c-numbers θ, η which *anticommute*, i.e.:

$$\theta\eta = -\eta\theta,\ \theta^2 = \eta^2 = 0.$$

It follows from this that functions $F(\theta)$ have a very simple form (consider the Taylor expansion!). Indeed, it is only integrals over sets of such numbers, the bispinor field configurations, that have physical significance. The anti-commuting properties of such numbers have the effect of interchanging certain determinants of operators in Gaussian integrals, and automatically accounting for the various sign conventions relating Feynman diagrams for fermion fields. If one is prepared to work with Grassmann field configurations as beables, there would again seem to be no problem in principle; the method just sketched can, in principle, be applied to this case as well.

5. The new problem of measurement

Here is a clear mathematical programme, but it is not so clear that we should embark on it. One objection is that the guidance equation, eqn 20, or eqn 22, is both non-local and non-covariant (the two claims are of course rather different). There has been plenty of discussion as to whether this will be an inevitable feature of any 'realist' solution to the problem of measurement. It has often been argued that the world, the phenomenology itself, is non-local, so that a theory adequate to the phenomenology had better be non-local as well. Some are prepared to conclude from this that we should not bother about covariance either, at least at the level of the beables, so long as we can demonstrate that, at the macroscopic level, some version of a 'no-signalling' theorem holds good (see, e.g., Maudlin, 1994). Similar claims are made on behalf of state-reduction theories.

But I will say nothing about these questions here, for there is a much more pressing problem. Call it the *new* problem of measurement: why is it that the field configurations are well-localized in space? And if they are not, in general, well-localized: what *special* considerations apply to suitable analogues of macroscopic bodies, to ensure that they are well-localized? If none, the measurement problem looms before us anew. It is no more use, here, to refer back to the non-relativistic limit, and to the particle beables that can be defined in that regime – local by definition of the concept of particle – than it would be to appeal to the *classical* limit, in the case of standard NRQM. Were we content with loose and formal arguments of that sort, there would have been no *old* problem of measurement, and no need to move to the pilot-wave theory in the first place.

An argument recently advanced is that indeed we can be sure that the field beables will be localized to within the non-relativistic wave-packet. It is due to Valentini (1992); he considered the real scalar field, adapting an argument to be found in Bohm et al. (1987). For simplicity, I shall state it in the non-relativistic case.

Given box-normalization, the ground state as a functional of complex normal modes is:

$$\Omega[a_1, \ldots, a_k, \ldots, a_1^*, \ldots, a_k^*, \ldots] = \exp\left(-\sum_k k a_k^* a_k\right). \tag{23}$$

We can use eqn 18 for the real scalar field as well. Inverting it we can write the c-numbers a_k, a_k^* in terms of the fields ψ, and then substitute in eqn 23. We obtain:

$$\Omega[\psi, t] = \exp\left(-\int\int \psi(\mathbf{x}, t)\psi(\mathbf{x}', t)f(\mathbf{x}-\mathbf{x}')\mathrm{d}^3\mathrm{x}\mathrm{d}^3\mathrm{x}'\right) \tag{24}$$

where

$$f(\mathbf{x}-\mathbf{x}') = \frac{1}{V^{1/2}}\sum_k k \exp(i\mathbf{k}.(\mathbf{x}-\mathbf{x}')).$$

In this way we can represent the vacuum state as a functional of real classical field configurations. We can extend this to n-particle states using the usual definition

$$|n>= \hat{a}_k^\dagger \ldots \hat{a}_{k'}^\dagger \ldots \hat{a}_{k''}^\dagger \ldots |0>$$

(n creation operators in all). Using eqn 23 to define the ground state, and representing the creation operators $\hat{a}_k^\dagger$ as derivatives, the 1-particle state of momentum $\mathbf{k}$ is:

$$\Psi_k(a_0, \ldots, a_k, \ldots] = a_k \exp\left(-\sum_k k a_k^* a_k\right).$$

Inverting eqn 18 once again we can represent this state as a functional of field-configurations ψ. For a 1-particle state of the form:

$$\varphi(\mathbf{x}, t) = \int f_k \exp(-i\mathbf{k}.\mathbf{x} - iE_k t)\mathrm{d}^3k$$

we find in this way:

$$\Psi_\varphi[\psi, t] = \int \varphi(\mathbf{x}, t)\psi(\mathbf{x}, t)\mathrm{d}^3x\Omega(\psi, t). \tag{25}$$

From this expression we can deduce that $\Psi_{\varphi}[., t]$ achieves its maximum value on configurations ψ whose support lies in the support of φ. It is obvious from eqn 25 that configurations ψ for which $\text{sup}(\varphi) \cap \text{sup}(\psi) = \emptyset$ have zero probability.

We may allow that there is a rigorous argument somewhere in this neighbourhood, but it is obvious that it will fall short of its purpose. For let it be granted that the 'most probable' classical field configuration can, in certain circumstances, be identified with the 1-particle non-relativistic state. From that it only follows that the problem of measurement will take the same form in pilot-wave RQFT that it takes in *standard* NRQM. What is needed is an argument to show that the classical field configuration will most likely be localized to regions *small* within the support of the non-relativistic state, in fact small on macroscopic length-scales, at least in measurement situations. That part of the field representing the apparatus pointer had better be localized in this sense. Otherwise we are no better off than in standard quantum mechanics.

How does the pilot-wave theory guarantee this result in the non-relativistic case? The answer is not reassuring: there is nothing specific to sub-systems with a large number of degrees of freedom (the macroscopic pointer), or to the dynamics, or to the low-energy limit; the locality of the beables is ensured by the simple proviso that they are point-like *by definition*, that they are *point particles* in the classical sense. If this is the method used in the non-relativistic case, what reason is there for thinking that classical field beables, which by definition are *not* point-like, will turn out to be localized where it counts? So far we have *no indication at all* that this will be so; *no reason at all* has been given to suppose that solutions to eqn 22 will, in appropriate circumstances, be localized to regions small on the macroscopic scale. This is the new problem of measurement.

Valentini's argument is so far from what is required that, on grounds of charity, we should ask whether he had some other point in mind. Indeed, it might seem that his aim was more modest:

> A realistic field description of particles may at first sight seem untenable . . . one may ask how a field distributed over all space can account for the highly localized massive particles seen in the laboratory. However, exactly the same query may be put to standard quantum field theory; for say the scalar case, the basic 'observable' is surely the field operator $\hat{\phi}(\mathbf{x}, t)$, whose eigenvalues are the set of definite field configurations $\phi(\mathbf{x})$, associated with eigenstates $|\phi(\mathbf{x})>$. How do 'particles' localized in space emerge? (Valentini, 1992, p. 50).

True enough, the standard field theory is no better in this regard; but then, for most of us, the *only* virtue of the pilot-wave theory is that it resolves the measurement problem. If it is not *better* than the standard theory in this respect, then it has nothing to commend it.

6. Whither hidden-variables?

Failing a demonstration that solutions to eqn 22 will, in the right circumstances, be localized – and no reason has been given to suppose that they will be – the pilot-wave theory had better make do with the mechanism it relies on in NRQM: the beables had better be particles. With that, whatever else might be wrong with the theory, there will be no problem of measurement.

Obviously it may be possible to come up with something entirely new. But as things stand nobody has; we had better consider again the options of section 3. Of the two considered, it would seem that the best strategy is Dirac's. What is needed is the 'negative-energy sea', and the restoration of a principled distinction between particles (fermions) and fields (bosons). In fact this fits much better with particle physics today, than it did almost fifty years ago, when Bohm first suggested that the pilot-wave theory should make use of Dirac's ideas. Now, but not then, we know that the constituents of ordinary matter are fermions, and that bosons are invariably associated with gauge fields (and hence with forces; the Higgs boson is the exception, but this is hardly a constituent of any ordinary matter). There is no cause for alarm if only fermions are sure to be localized, for they are all that we ordinarily see.

Yet the hole theory has few advocates among those who support the pilot-wave theory. Thus Valentini (speaking of QED): 'Not only is the (hole) theory very inelegant, it also creates an unsatisfactory dualism: particle description for the massive case versus field description for the massless case' (ibid., p. 50.) And thus Bohm, Hiley, and Kaloyerou: 'Although this theory provides a consistent interpretation, it is somewhat ad hoc and consequently is not likely to provide a great deal of further insight' (Bohm et al., 1987, p. 374.) So whither hidden variables in relativistic quantum physics? And whither hidden variables in the *non-relativistic* case, if there is no such insight to be had in the relativistic domain? But according to Bohm et al., it seems that this is not a problem:

> However, it is our view that, at this stage, it is premature to put too much emphasis on the interpretation of relativistic quantum mechanics. This is because we feel that the theory in its current stage of development is probably not consistent enough to be given an overall coherent interpretation. First of all, there are the infinities which make it difficult even to see what the theory means. For example, the dressed particles are said to be in a different Hilbert space from which the theory starts. (Ibid., p. 374)

The Copenhagen interpretation has long been a refuge for obscurantists; it would be a pity if advocates of the pilot-wave theory were to resort to similar evasions. There are mathematical problems aplenty in renormalization theory, but that should not be allowed to hide the fact that relativistic quantum physics, in the case of special relativity, is a detailed and successful physical theory, as precise and systematic as anything in physics (more so; but the lesser point is enough). It is no more 'premature' to seek to put this theory on a principled footing than it is to put NRQM on a principled footing. If, notwithstanding the remarkable success of the standard model, nothing less than a full-blown theory of quantum gravity will do (the quoted passage goes on to cite difficulties with reconciling quantum theory and *general* relativity), it is hard to see why we should bother with NRQM in the first place – hence neither with the pilot-wave theory. There is an honourable tradition, well known through Penrose's writings (Penrose 1989), according to which the problem of measurement is to be resolved by a proper marriage of quantum theory with gravity; but that is not what advocates of the de Broglie–Bohm theory are appealing to.

Failing something radically new, or a proof that eqn 22 yields localized field configurations, we are left with particles as beables, and with the approaches of Dirac and Feynman. I have already indicated my own opinion, that Dirac's offers the better prospects for the pilot-wave theory, but in truth both have their drawbacks. The hole theory is *not*, contrary to what is sometimes claimed (Bohm & Hiley, 1993, p. 276), unitarily equivalent to conventional QED. It is simply not true that 'in most cases the wave function will factorize, so that it will be sufficient to consider a limited number of particles and ignore the rest' (Bohm et al., 1987, p. 375), or that the difficulty is 'similar . . . to what happens with the boson fields for which likewise the whole universe must be considered in principle, while in practice a limited number of Fock states may be adequate' (ibid., p. 375). On the contrary, in any pair creation or annihilation process the negative energy sea had better play a dynamical role. Any calculation of the probability amplitudes will have to take into account the infinite number of electrons already *in situ*. Since there is no unitary mapping between the finite-particle states of conventional QED, and the infinite-particle states of the hole theory, we must work from the outset with particle interactions, and with the negative-energy sea.

There is, however, an historical precedent: progress, albeit limited, was made with the hole theory in the early 1930s. But there is a final consideration, that may yet tell in favour of Feynman's approach. For at bottom the pilot-wave theory is only of interest insofar as it provides a *believable* interpretation. Unlike the original Dirac hole theory, whose interest lay for the most part in what it had to say about new phenomena, and in the combining of principles of relativity theory and quantum mechanics, the virtue of the pilot-wave theory is that it solves the problem of measurement. But can we really *believe* in the literal existence of an infinity of point-particles, in the classical sense, all with negative energy and charge, in every non-zero volume of space? To paraphase Putnam, speaking of a very different interpretation: what is the point of an interpretation of quantum mechanics that one cannot *believe*? (Putnam 1990, p. 10). But that was not Dirac's view. He, along with most others of the time, took an instrumentalist view of the theory. Only experiment and questions of elegance and mathematical simplicity really mattered to him. It was in that climate of opinion that the hole theory was seriously entertained. That is hardly an outlook available to the pilot-wave theory.

References

Berezin, F. (1996). *The Method of Second Quantization* (New York: Academic Press).

Bohm, D. & B. J. Hiley (1987). 'I. Non-relativistic Particle Systems', *Physics Reports* **144**, 323–48.

(1993). *The Undivided Universe* (London: Routledge).

Bohm, D., B. J. Hiley & P. N. Kaloyerou (1987). 'II. A Causal Interpretation of Quantum Fields', *Physics Reports* **144**, 349–75.

Bohm, D., R. Schiller & J. Tiomno (1955). *Suppl. Nuovo Cimento* **1**, 48–66.

Holland, P. (1993). *The Quantum Theory of Motion* (Cambridge: Cambridge University Press).

Kypianidis, A. (1985). *Physics Letters A* **111**, 111–16.

Maudlin, T. (1994). *Quantum Non-Locality and Relativity* (Oxford: Blackwell).

Penrose, R. (1989). *The Emperor's New Mind* (Oxford: Oxford University Press).

Putnam, H. (1990). *Realism With a Human Face* (Cambridge, MA: Harvard University Press).

Saunders, S. (1991). 'The Negative Energy Sea', in S. Saunders and H. Brown (eds.), *Philosophy of Vacuum* (Oxford: Clarendon Press).

Valentini, A. (1992). 'On the Pilot-Wave Theory of Classical, Quantum and Sub-Quantum Physics', PhD thesis, International School for Advanced Studies, Trieste.

5

Bohmian mechanics and chaos

JAMES T. CUSHING AND GARY E. BOWMAN

1. Introduction

It is a pleasure to be able to contribute to this volume in honour of Professor Michael Redhead. In the foundations of physics community, his influence has been felt both through his *Incompleteness, Nonlocality and Realism* (1987), which has largely defined the terms of discourse in this field for the last decade, and by the new 'Cambridge School' of the philosophy of physics that he established there during his Professorship in the History and Philosophy of Science Department. Michael Redhead's work illustrates well what has always seemed to be a central tenet of his: that philosophers of science should use the conceptual background and the mathematical formalism of modern, successful scientific theories – especially quantum mechanics and quantum field theory – as a guide to understanding the ontology of the world (Redhead 1983, 1990) and to reexamining traditional philosophical questions (French & Redhead 1988; Redhead & Teller 1991, 1992). In his Tarner Lectures (1995) he used this type of work to anchor a general position that can be characterized as *structural realism*, which holds that the structures – mathematical formalism, models, analogies – of a successful scientific theory eventually define the real, even as regards central, unobservable entities.

While Professor Redhead has done an outstanding job of pressing forward with this general programme, it remains arguable that he and some of his co-workers, who have, in the main, stayed within the framework of standard quantum theory and relativity, may have effectively bracketed a different and potentially fruitful alternative theory. In this essay we attempt to illustrate how a prior commitment to standard interpretations (say, of relativity or of quantum mechanics) can effectively predetermine the answer to certain foundational philosophical questions. In fact, the conclusions drawn from philosophical analyses based on traditional interpretations of these theories may be completely inverted when one rehearses those same arguments, but beginning with an alternative interpretation.[1] Then we see with a vengeance the indeterminateness of responses to these basic questions that results from the underdetermination of two empirically equivalent scientific theories.[2]

2. A choice of Bohm versus Copenhagen[3]

Since the Copenhagen and Bohm versions of quantum mechanics are generally taken to be empirically equivalent, we begin by considering how one might reasonably choose between these two theories in a non-question-begging fashion.

If one feels that quantum mechanics should provide the basis for an acceptable ontology, then one could object that Copenhagen does not do this. That is, although it is sometimes claimed that Copenhagen quantum mechanics has a 'wave' ontology (Bedard 1997), it is, nevertheless, not obvious that this can be so.[4] Certainly, such a putative ontology may at first sight seem plausible for a single 'particle' or wave existing in a three-dimensional space (where configuration space could be confused with ordinary three-dimensional physical space), when it might be thought that a 'narrow' wave packet would represent a (classical) particle. But such an interpretation becomes problematic for a system of two or more particles, since the (Schrödinger) waves then exist in a multidimensional configuration space. It is difficult to see how such waves – even if narrow – could represent a collection of particles, because a collection of N particles exists in physical three-dimensional space, whereas the corresponding waves exist in a $3N$-dimensional configuration space.[5] This, of course, is basically a difficulty

[1] These questions are treated at length in Belousek (1998). An example is whether or not one can, in principle, distinguish from each other, by their various histories (or trajectories), what are usually taken to be 'identical' particles in quantum mechanics.

[2] An extensive discussion of the underdetermination thesis in the context of the Copenhagen versus the Bohmian versions of quantum mechanics can be found in Cushing (1994).

[3] Although one can argue that Louis de Broglie already in 1927 had much the same theory that David Bohm rediscovered in 1952, for convenience in this essay we refer simply to 'the Bohm theory' or to 'Bohmian mechanics', rather than to 'the de Broglie–Bohm theory'.

[4] In fact, some of the early, central figures who shaped the various 'Copenhagen' interpretations of quantum mechanics supported a particle ontology (cf. Beller (1990), especially pp. 575–6).

[5] Admittedly, one could attempt to argue that, perhaps after all, we actually do exist in a higher dimensional configuration space (Albert 1996). However, we do not pursue this possibility here.

that Schrödinger had with his own early attempt to base a wave-only ontology on the wave function ψ.[6]

However, this is simply a criticism of a standard interpretation of quantum mechanics, one that might be given some weight by an opponent of Copenhagen, but that is unlikely to dissuade a supporter from his or her view on this matter. Similarly, one might look for a certain lack of coherence (but not strict logical inconsistency) in Bohmian mechanics (BM). Such an attempt was the claim that where a detector registers a Bohmian particle to be can be different from the location of the trajectory on which the theory implies it should be (Englert et al. 1992). Upon further analysis, though, this criticism turns out to be less telling than it appears at first sight.[7] So, having tried some negative ploys against each of our candidates, let us consider a more positive approach.

The first and most obvious possibility is that these two theories are, after all, *not* completely equivalent in their predictions.[8] For instance, Valentini (1991b, 1992) has exploited the fact that, in Bohm's theory, the basic conceptual role of the wave function ψ is a dynamical one of guiding the motion of the particle, so that there is no conceptual necessity within BM that the probability density P must be equal to $|\psi|^2$. (The condition $P = |\psi|^2$ is often referred to as *quantum equilibrium*.) Specifically, if the distribution of matter in the early universe was not in quantum equilibrium, and yet that distribution evolved into quantum equilibrium (Valentini 1991a, 1992), then we might look for cosmological traces of this early disequilibrium. If this were to turn out to be the case, then the decision between Bohm and Copenhagen would be of the familiar type based on different predictions made by each. Let us be somewhat pessimistic, assume that this will not happen, and press on.

Next, one might hope that, even though these two theories do not disagree on any predictions that they are both able to effect in common, there would be predictions for some class of phenomena that one theory could make, while the other is unable to make any at all. While it has been known for some time that BM can consistently define particle tunnelling times in the absence of any measurement (when Copenhagen cannot),[9] the question

[6] It is a *very* different matter (as in BM) to have in three-space N actual particles that are guided by a wave in a $3N$-dimensional configuration space.

[7] The history of this development can be followed in Englert et al. (1992, 1993), Dürr et al. (1993), Dewdney et al. (1993), Aharonov & Vaidman (1996) and Barrett (1997, 1999).

[8] See Becker (1997) for an analysis of whether or not a 'no-collapse' theory such as Bohm's can in principle be distinguished from the standard 'collapse' (or projection postulate) interpretation of quantum mechanics.

[9] While there is a huge literature on this subject, Leavens & Aers (1993), Cushing (1995) and Challinor et al. (1997) together provide an overview of the situation on this question.

remains whether or not these tunnelling times (like the Bohmian trajectories themselves) remain unobservable in principle. There have been suggestions that quantum tunnelling times through a barrier might provide such an opportunity (Cushing 1995). However, this does not seem to be possible (Bedard 1997), the basic reason being that in BM all accessible information is configurationally based (Dürr et al. 1992a). That is, for Bohm all measurements are ultimately position measurements and the distribution of these results must be given by $|\psi|^2$, both in Bohm and in Copenhagen. Hence, this does not appear to be a promising avenue for the resolution of our underdetermination.

We find ourselves, then, reduced to new insights and new avenues of research that could be suggested by Bohm's formulation. Into the former category would fall the well-known absence of the measurement problem (Bohm 1952a; Dürr et al. 1992a; Cushing 1994) in BM, progress toward a coherent treatment of the classical limit (Holland 1993, 1996) and the meaning of the numerical output of a tunnelling-time experiment (Cushing 1997). The price tag for all of this is non-locality. But such non-locality is arguably unavoidable in any empirically adequate quantum theory (Maudlin 1994; Clifton and Dickson 1998), so that these features of BM might be seen as net gains over Copenhagen. Furthermore, such a move also makes reasonable a possibly very fruitful reexamination of the basis for the common acceptance of special relativity as a meta-principle, rather than simply as an empirically grounded constraint on observations and predictions (Selleri, 1994; Cushing 1996; Valentini 1997). The motivation there would be an attempt to resolve the tension between a realistic interpretation of quantum mechanics and the usual gloss put on the commitments necessitated by special relativity. Basically, this requires for special relativity critical foundational analyses of the type that abound in the 'Bell' literature about non-locality and quantum mechanics.[10] Into the latter category – that of new avenues of research – fall recent attempts to apply Bohmian field theory (BFT)[11] to quantum gravity (Valentini 1992; Vink 1992; Horiguchi 1994; Callender & Weingard 1994; Shtanov 1996; Blaut & Glikman 1996). Here the hope is access to new physics, rather than any direct disagreement between the two theories.

[10] Here Jarrett (1984) and Cushing & McMullin (1989) can serve as a representative sampling of this huge literature.

[11] Valentini (1992, 1996, 1999) has provided, in the relativistic domain, a Bohmian formulation of quantum field theory (QFT), one whose statistical predictions are identical to those of standard relativistic QFT.

3. A synopsis of Bohmian mechanics

Before we can discuss the possible relation between chaos in classical mechanics (CM) and quantum chaos (in BM), we need a precise framework within which to pose a proper mathematical problem.[12] So, let us turn to an outline of BM and a consideration of how one obtains a CM regime from BM. We can then sketch a comparison between the possibilities for chaos in CM and in BM for the three classes of interest: chaos in CM, but none in BM; no chaos in BM, but chaos in CM; and chaos in CM and in BM.

Here we simply give a summary of the analytical machinery available in BM for studying the trajectories of particles (Dürr et al. 1992a; Holland 1993; Cushing 1994; Berndl et al. 1995). If we begin with the Schrödinger equation

$$i\,\hbar(\partial\psi/\partial t) = H\psi = (T+V)\psi = -(\hbar^2/2m)\nabla^2\psi + V\psi \tag{1}$$

where $V = V(\boldsymbol{x})$, and make the polar decomposition of $\psi = Re^{iS/\hbar}$ suggested by Bohm (1952a), then we obtain a set of coupled equations for R and S

$$(1/2m)(\nabla S)^2 + V + U = -\partial S/\partial t \tag{2}$$

$$\partial P/\partial t + \nabla\cdot[(P/m)\nabla S] = 0 \tag{3}$$

where $P = R^2 = |\psi|^2$ and the so-called *quantum potential* $U(\mathbf{x}, t)$ has the form $U(\boldsymbol{x}, t) \equiv -(\hbar^2\nabla^2 R)/(2mR)$. If we treat eqn 2 as a (modified) Hamilton-Jacobi (H-J) equation and take (as an *Ansatz*) $\boldsymbol{p} = \nabla S$, so that the corresponding Hamiltonian would be $H = (\boldsymbol{p}^2/2m) + V + U$, then we find (from $\dot{\boldsymbol{x}} = \partial H/\partial\boldsymbol{p}$) the 'guidance condition'

$$\mathbf{v} = \dot{\boldsymbol{x}} = \boldsymbol{p}/m = (1/m)\nabla S(\boldsymbol{x}, t)|_{\boldsymbol{x}=\boldsymbol{x}(t)} \tag{4}$$

which gives a velocity field $\mathbf{v}$ for the trajectories $\boldsymbol{x}(t)$. This is the velocity $\mathbf{v}$ that enters into the continuity equation, so that eqn 3 can be rewritten as

$$\partial P/\partial t + \nabla\cdot(P\mathbf{v}) = 0 \tag{5}$$

This velocity $\mathbf{v}$ is, in fact, simply related to the usual quantum-mechanical current density $\boldsymbol{j} = (i\hbar/2m)(\psi\nabla\psi^* - \psi^*\nabla\psi)$ as $\mathbf{v} = j/P = (1/m)\nabla S$. One often writes $E = -(\partial S/\partial t) = T + V + U$, so that the quantum potential U

[12] See Belot & Earman (1997) for a discussion of this question within the framework of a more standard formulation of classical and quantum mechanics.

is a new form of energy. However, the 'energy' E thus defined is not necessarily a constant, because ψ can have explicit time dependence.

Notice that ψ, R and S all depend upon $\hbar$ (and this is the source of the trouble in using anything like the WKB approximation to define the classical domain as some type of limit as $\hbar \to 0$). Furthermore, R and S now mutually determine each other (unlike the situation that obtains in CM if we use the H-J equation to define an ensemble of possible trajectories). Because of the continuity conditions imposed on the wave function ψ, ∇S is single-valued away from nodes (i.e., places where $R = 0$) and trajectories cannot pass through nodes. In order to determine ψ uniquely, one must specify $\psi_o(\boldsymbol{x}) = \psi(\boldsymbol{x}, t = 0)$ over all space initially (assuming that ψ and $\nabla\psi$ vanish at spatial infinity, as is typically the case). If we were to treat eqn 2 as a modified (or 'quantum') H-J equation for a function S_{HJ}^{QM}, then a complete integral would depend upon a set of non-additive constants of the motion, just as in CM (Goldstein 1950, ch. 9). In BM one postulates (Bohm 1952b) $S_{BM} = S$ (i.e., the phase of the wave function ψ), but one does not derive this as a general result from the 'quantum' H-J equation. That is, the assumption (or 'guidance condition') of eqn 4 is verified to give *a* solution to the equations of motion. These solutions for $S_{BM} = S$ are only a subset of all of the solutions that would be contained in S_{HJ}^{QM}. This observation is important for what follows since, in general, this larger set of solutions will correspond to ensembles for which P is different from $|\psi|^2$ (and, hence, would not be empirically adequate). That is, such a more general class of solutions to the 'quantum' H-J equation would give statistical predictions that disagree with those of standard quantum mechanics (Holland 1993; Baublitz & Shimony 1996).

It is a straightforward exercise to show that the total time derivative of the velocity field of eqn 4 yields simply

$$\frac{d\boldsymbol{p}}{dt} = -\nabla(V + U) \tag{6}$$

which is just Newton's second law of motion, but now with the additional quantum-potential term. The Bohmian trajectories obtained from the first-order eqn 4 never cross (in configuration space). In BM, unlike CM, specifying the values of V, $\boldsymbol{x}_o$, $\boldsymbol{p}_o$ (really, just V and $\boldsymbol{x}_o$, because $\boldsymbol{p}_o = \nabla S(\boldsymbol{x}_o)$) does not uniquely determine a trajectory, since we do not have a specific dynamics or solution until the quantum state ψ is also given. This simply reflects the fact that in BM a complete specification of the state of a system is $(\psi, \boldsymbol{q})$, not just ψ alone.

4. The classical domain[13]

A classical regime should emerge from BM, not be postulated as a separate entity. Some of the common standard suggestions for obtaining a 'classical limit' of quantum mechanics have been (i) $\hbar \to 0$, (ii) n (some quantum number) $\to \infty$, (iii) λ (some wavelength) $\to 0$ or (iv) m (mass) $\to \infty$. All of these are problematic and exceptions can be found (Holland 1993, ch. 6; Ford & Mantica 1992).

In light of eqn 6, we take as the criterion for the classical domain the requirement that we obtain the equations of CM from those of BM (namely, $dp_j/dt = -(\partial V/\partial q_j)$, $dE/dt = (\partial V/\partial t)$, which follow from eqn 6 and 4, once $U \equiv 0$).[14] That is, we demand that $U \to 0$ in the sense that it be negligible compared to other relevant energies in a given problem. This is the *correspondence principle* of Bohmian mechanics (Holland 1993) and is what we will mean by a classical limit. The condition $U \to 0$ is state dependent (i.e., it depends on the wave function ψ for the system) so that not all quantum-mechanical systems have classical limits. On the other hand, that $S_{BM} \equiv S \to S_{limit}^{CM}$ will not necessarily be a complete integral means that to an arbitrary S_{HJ}^{CM} there need not, and usually does not, correspond some S_{BM} from which it can be obtained. One particularly simple way to see this is that (for a given pure state ψ) in BM each given initial position $\boldsymbol{x}_o$ has just *one* possible momentum $\boldsymbol{p}_o$ (or velocity $\mathbf{v}_0$) associated with it (cf. eqn 4). On the other hand, in CM $\boldsymbol{x}_o$ and $\boldsymbol{p}_o$ are independently assignable.

The use of mixed states allows one, for a given $\boldsymbol{x}_o$, to have $\boldsymbol{p}_o$ take on various values for different sub-ensembles. However, this does not, in itself, take care of a more serious objection based on the non-crossing property of the set of all trajectories corresponding to a given *pure* state. Holland (1996) has a nice example of the trajectory pattern for BM versus that for CM when one considers an ensemble of particles, each particle beginning from a different initial position, but all moving (parallel to each other) toward a perfectly rigid wall. These particles (or trajectories) are then reflected from the wall. For CM, the pattern of reflected trajectories has those trajectories crossing incoming ones, while for BM different trajectories can never cross each other. Therefore, one could not, by any limiting process, obtain the CM pattern from the BM one.

[13] An excellent discussion of the classical limit in BM is given in Holland (1993, ch. 6, and 1996). Holland takes the position that BM is unlikely to be able to underwrite all of CM.

[14] If we require that the wave function be a narrow packet, then all of the Bohmian trajectories are 'crowded' together, implying that a sharp, observable, essentially unique trajectory will be followed.

Could there be any way to avoid this killing objection? One possibility is the following, which is by no means obviously doable. Suppose – and this remains to be argued for – that (for a given potential V) one could construct enough linearly independent 'classical' states ψ^c for which the quantum potential (U) is essentially zero. Then, for any *given* initial values $\boldsymbol{x}_o$ and $\boldsymbol{p}_o$ there could be (at least) one ψ^c for which the BM trajectory would be the CM one. Let us write such a state as $\psi^c(\boldsymbol{x}_o, \boldsymbol{p}_o)$. Specifically, if $\psi^c(\boldsymbol{x}_o, \boldsymbol{p}_o)$ has almost zero support except around $\boldsymbol{x}_o$, then this BM system would follow a CM trajectory.[15] A mixed state made up of these various $\psi^c(\boldsymbol{x}_o, \boldsymbol{p}_o)$, would allow one effectively to mimic, or generate, the pattern of CM trajectories, each trajectory corresponding to a given $\psi^c(\boldsymbol{x}_o, \boldsymbol{p}_o)$.

Settling these questions one way or the other should help clarify whether or not it is reasonable (or even possible) to underwrite *all* of CM with BM. If such a construction is not possible, then, in addition to (i) quantum systems with no classical analogue, there could be (ii) classical systems that are not the limit of any quantum system. The latter would not be good news for those who would like to consider BM as the fundamental theory in both the classical and quantum domains.

5. Quantum chaos

Even though it has long been expected (Bohm 1952a) that the trajectories of BM could exhibit chaotic behaviour (roughly, initially adjacent trajectories in phase space that subsequently diverge very rapidly in time[16]), it has been only recently (Parmenter & Valentine 1995, 1997; Faisal & Schwengelbeck 1995; Guglielmo & Pettini 1996; Frisk 1997) that specific cases of such motion have been explicitly demonstrated in the literature. As we discuss below, one of the implications of these results could be that chaotic motion in CM is neither necessary nor sufficient for the possibility of chaotic motion in BM (i.e., quantum mechanically). This raises an interesting technical and conceptual question of how chaotic motion at the quantum (BM) level survives, emerges from or disappears in the limit to the classical domain (when such a limit exists).

15 This is, in fact, very much what happens in the $U = 0$ limit of the simple harmonic oscillator. It is also what one might naively expect from Ehrenfest's theorem for a highly localized state. Observations like these leave aside the question of under what physical conditions such states are feasible.

16 For a precise characterization of chaotic motion in CM see, for example, Guckenheimer & Homes (1983) or Rasband (1990).

Our motivation for considering this problem is that useful insights into quantum chaos may follow from Bohm's programme. A basic difficulty for standard quantum theory is that of specifying just what quantum chaos is because the typical hallmark of chaos is[17] the highly irregular behaviour of the trajectories of a system. In the quantum domain, there still exist trajectories for Bohm, but not for Copenhagen. Because of this lack of a conceptual framework within which to formulate the question of quantum chaos in standard quantum mechanics, there have been suggestions that an essentially new domain, that of 'semiclassical' mechanics, be defined and explored and, possibly, even be provided with its own ontology (Rohrlich 1988; Berry 1989; Batterman 1991, 1993). Batterman's excellent article (1991) focuses on what we, like him, take to be the central issue for the question of defining quantum chaos: the relation between quantum mechanics and classical mechanics. Since there is in standard quantum mechanics no well defined limit in which classical mechanics emerges from quantum mechanics, Batterman suggests reevaluating the meaning of Bohr's correspondence principle in terms of a semiclassical ('old quantum theory') regime that straddles and in some way connects the quantum and the classical worlds and to look for signatures of chaos in this semiclassical regime as a clue to quantum chaos. Berry (1987, 1989) has studied quantum chaology in this semiclassical regime.[18]

Related to this programme are several attempts to characterize (standard) quantum-mechanical chaos in terms of the time evolution of expectation values, the time dependence of the overlap of two state vectors initially identical but evolving according to Hamiltonians with slightly different control parameters, energy level crossing and repulsion (Haake 1991), energy level spacing distributions (Tabor 1989) and the 'scarring' of wave functions for highly excited states (Gutzwiller 1990; Wilkinson et al. 1996).[19] As a result of examining many particular systems, there have emerged correspondences, or rules of thumb, that indicate when one may expect certain types of 'quantum chaos' behaviour, based on the regular or chaotic behaviour of the corresponding classical system. Still, there does not at present seem to be any general understanding of why these various 'signatures' (really, essentially just somewhat arbitrary candidates for definitions or criteria) of quantum chaos should be connected and why they should be related to the corresponding classical behaviour of the system.

[17] Deutsch (1997, p. 202) makes the dramatic move of denying the existence of actual physical chaos because he takes QM to be the basic theory and there is (according to him) no chaos in QM. However, this position is tied to his parallel-universes interpretation of QM.

[18] For an introduction to this semiclassical programme, see Tabor (1989, ch. 6).

[19] A readable and concise summary of these criteria is given in Haake (1991, pp. 1–11).

It is not evident how to find some common feature of classical and quantum systems that can characterize chaos. One reason for this, as Batterman (1991, p. 196) puts it, is that: 'Another way of seeing that QM is inhospitable to chaos is by recognizing that in QM there is no analog to the classical phase space trajectory'. However, the fact that there are trajectories in Bohm's (observationally equivalent) theory can provide a conceptual matrix within which to characterize and discuss 'quantum chaos' in terms of the standard criteria employed in CM (Cushing 1994, pp. 157 and 192). From a Bohmian perspective, there may be no need for moves of the type made by Berry and by Batterman, because one ontology and one dynamics would then govern both the quantum (BM) and the classical domains. Detailed study of simple (Bohmian) mechanical systems that can be treated with a combination of analytic and numerical techniques (Parmenter & Valentine 1995, 1997; Faisal & Schwengelbeck 1995) may prove quite rewarding in contributing to such an insight.

Recently, there have also been suggestions (Schack & Caves 1993; Schack et al. 1994; Caves & Schack 1997) that quantum chaos be defined in terms of the amount of information needed to follow the evolution of a perturbed vector in Hilbert space (i.e., the distribution of vectors into which the initial vector may evolve). One can then attempt to characterize both classical and quantum chaos in this fashion in order to give a unified description of chaos. Even if this programme, within the framework of standard quantum mechanics, should prove to be a generally reliable indicator of quantum chaos, it is unclear that this would give us a genuine understanding of just what quantum chaos is. However, we might also exploit the resource of trajectories in BM to put quantum chaos on a common footing with classical chaos[20]. This is the approach we consider now.

6. Chaos in Bohmian mechanics

Let us begin with three examples of systems that have been examined for chaos in BM and that give generically different results. First, there is a

[20] Schwengelbeck & Faisal (1995, p. 283) do precisely this when they define a Kolmogorov–Sinai (KS) entropy (a measure of stochasticity) for Bohmian trajectories and take this quantum dynamics to be 'chaotic, if in a given region of phase space the flow of trajectories, according to the Hamilton–Jacobi formulation of quantum mechanics, has positive KS entropy'. Iacomelli and Pettini (1996) and Sengupta and Chattaraj (1996) also attempt to use BM to relate classical and quantum chaos. There still remains the problem, though, of understanding chaos in CM in terms of that in BM.

system (the Arnold cat map) that is chaotic both in CM (Arnold & Avez 1968, pp. 5–9; McCauley 1993, pp. 114–15; Nakamura 1993, pp. 193–5) and in BM (Faisal & Schwengelbeck 1995).

Next, we can ask whether or not an isolated BM system left to itself can develop chaotic motion (as in some CM examples of deterministic chaos). The *Poincaré–Bendixson theorem* (Hirsch & Smale 1974, p. 248) insures that there can be no chaotic motion for so-called planar-flow systems (e.g., for mapping problems with just *two* variables, say x and t).[21] Hence, we know that the corresponding BM system will need at least two dependent variables (in addition to the time t). Parmenter & Valentine (1995) have studied (numerically) such a case – the anisotropic harmonic oscillator in two dimensions.[22] They found that chaos (for the trajectories in configuration space) can be exhibited, provided that the wave function $\Psi(x, y, t)$ for the system is a superposition of at least three energy eigenstates, with two of the corresponding frequencies being incommensurate. (The last condition prevents the wave function, and hence the guidance condition, from being periodic in time.) This system is not chaotic in CM. In fact, because the Hamiltonian for this system is additive in the dependent variables (x and y), one can argue that any state ψ for which $U \equiv 0$ will factorize into an unentangled state, thus yielding two uncoupled systems, each of which is known not to be chaotic.[23] So, for this example, any classical-limit state that yields $U \equiv 0$ can have no chaotic behaviour. The 'classical limit' of this system could not exhibit chaos and there would be no disagreement between BM and CM here.

Finally, the kicked rotator is a paradigm of driven chaotic motion in one (angular) dimension in CM (e.g., Lichtenberg & Lieberman 1992) and has been studied extensively even in conventional QM (Casati et al. 1979; Casati 1985; Izrailev 1990; Chirikov 1991). Classically, the kicked rotator depends on just one critical parameter K, such that this system executes regular motion for small values of this parameter and becomes chaotic for large values. In the quantum-mechanical case (BM) (Malik & Dewdney 1994; Schwengelbeck & Faisal 1995), it appears that there may be no chaotic behaviour for any *pure* state ψ.[24]

[21] Of course, this would cover only the trajectories in (this one-dimensional) configuration space, not necessarily in the corresponding (two-dimensional) phase space.

[22] They actually establish chaotic behaviour in a two-dimensional configuration space, but that is good enough to imply chaotic behaviour in the corresponding four-dimensional phase space.

[23] That is, once $U \equiv 0$, then eqn 6 is just the second-order (in time) version of Newton's second law of motion and our problem then becomes separable. However, the solutions to eqn 4 must also be solutions to eqn 6 (but not *vice versa*).

[24] If one considers mixed states, then trajectories of the ensemble can cross, so that chaos may still be possible here.

If, in fact, there are cases like this last one (for which there can be no chaotic motion in BM, and yet there is in CM), then that could present a conceptual difficulty for any attempt to underpin CM with BM. To overcome this, it may be essential to take into account the interaction of the BM system with its environment.

For the kicked rotator problem with measurements (Malik & Dewdney 1994), there is the possibility of chaotic behaviour.[25] With such an open quantum system it is, perhaps, not amazing that chaotic behaviour can be induced.[26] That is, initially closely neighbouring (BM) trajectories can become widely separated as a result of encountering 'bifurcation' points at these measurement interventions, after which they have been channelled onto rapidly diverging paths. Just as 'measurements' can be important here for obtaining expected classical behaviour from a quantum system, so, even within the framework of standard quantum mechanics, a necessary ingredient for the transition from the quantum domain to the classical one is often taken to be the interaction of the quantum system with the environment. For an open system, this can produce what is effectively the reduction of a pure state to a mixed state.[27] The classical limit is intimately involved in our understanding the connection between chaos in BM and in CM. And, interaction with the environment seems essential for classicality. It is not unreasonable, then, to examine the role that such interaction may play in an explanation of how CM chaos can emerge from a nonchaotic BM system.[28]

The technical aspects of this question remain considerable. Given our discussion in section 4 and in this one, we now see that both mixed states and interaction with the environment may be relevant for the possibility of such behaviour.

7. Conclusions

There are, it seems to us, two interesting general philosophical points that emerge from this discussion. First, we see that BM has certain conceptual

[25] Dewdney and Malik (1996) give a similar discussion for the quantum pinball with measurements.

[26] See Dürr et al. (1992b), Spiller & Ralph (1994), Brun (1995) and Dewdney & Malik (1996) for examples of chaotic behaviour in such open quantum systems. This type of chaotic behaviour is to be distinguished from Hamiltonian chaos (i.e., chaotic behaviour for an isolated system).

[27] On such environmentally induced decoherence, see Zurek (1991, 1993), and on its relation to chaos, see Brun (1995) (and references therein).

[28] In fact, these various approaches to obtaining a classical regime from quantum mechanics may be converging to a common set of explanatory resources.

and technical resources (not available in standard QM) that may allow it to address, and arguably to resolve, two long-standing and difficult issues in quantum theory (i.e., the classical limit and quantum chaos).[29] The BM representation of quantum phenomena allows one to state a precise condition – the correspondence principle of BM (i.e., $U = 0$) under which equations valid in the quantum domain pass over into the equations of CM. This formulation of the basic equations of motion also shows how it may be possible for the chaotic behaviour of classical trajectories to emerge from the chaotic motion present in BM. This would allow a unified (mathematical and conceptual) treatment of chaos in the classical and quantum domains.

Secondly, and even more significant in its overarching implications for quantum theory, there might be some (mathematical) classical systems that cannot be reached as the limit of a BM system. Then, the ubiquity of chaos in classical mechanics, as opposed to the situation in quantum mechanics, could turn out to speak against quantum mechanics (i.e., BM) being the fundamental theory of physical phenomena. Rather than classical mechanics being included in quantum mechanics, Bohmian mechanics and classical mechanics might simply possess a common region of (non-inclusive) overlap. In that case, neither classical mechanics alone[30] nor quantum mechanics alone would be sufficient, even in principle, to account for all of the physical phenomena we encounter – unless, of course, not all mathematically expressible classical systems are actually realizable in nature, but only those obtainable from BM. It would probably appear more plausible that CM and QM (BM) each have domains of validity excluded by the other theory. If so, neither theory would be fundamental and we would then have to look for a broader, more basic theory from which both CM and BM could be obtained in suitable limits. However, the jury is still out on this.

References

Aharonov, Y. & L. Vaidman (1996). 'About Position Measurements Which Do Not Show the Bohmian Particle Position' in Cushing et al. (eds.)(1996), pp. 141–54.

Albert, D. Z. (1996). 'Elementary Quantum Metaphysics' in Cushing et al. (eds.) (1996), pp. 277–84.

[29] Recall that BM long ago dispensed with another central difficulty of quantum mechanics – the measurement problem (Bohm 1952a; Dürr et al. 1992a; Holland 1993; Cushing 1994).

[30] Ford (1988, 1989), unlike Deutsch (1997), has expressed reservations about the 'completeness' of QM.

Arnold, V. I. & A. Avez (1968). *Ergodic Problems of Classical Mechanics* (New York: W. A. Benjamin).

Asquith, P. D. & T. Nickles (eds.) (1983). *PSA 1982*, Vol. II (East Lansing, MI: Philosophy of Science Association).

Bai-Lin, H. (ed.) (1988). *Directions in Chaos*, Vol. 2 (Teaneck, NJ: World Scientific Publishing Co.).

Barone, M. and F. Selleri (eds.) (1994). *Frontiers of Fundamental Physics* (New York: Plenum Press).

Barrett, J. (1997). 'The Persistence of Memory: Surreal Trajectories in Bohm's Theory', University of California, Irvine, preprint.

(1999). *The Quantum Mechanics of Minds and Worlds* (Oxford: Oxford University Press).

Batterman, R. W. (1991). 'Chaos, Quantization, and the Correspondence Principle', *Synthese* **89**, 189–227.

(1993). 'Quantum Chaos and Semiclassical Mechanics' in Hull et al. (eds.) (1993), pp. 50–65.

Baublitz, M. & A. Shimony (1996). 'Tension in Bohm's Interpretation of Quantum Mechanics' in Cushing et al. (eds.) (1996), pp. 251–64.

Becker, L. (1997). 'Testing for the Time of Collapse in Quantum Mechanics', unpublished PhD dissertation, University of Illinois at Chicago.

Bedard, K. (1997). 'Reply to "Quantum Tunneling Times: A Crucial Test for the Causal Program?"', *Foundations of Physics Letters* **10**, 183–7.

Beller, M. (1990). 'Born's Probabilistic Interpretation: A Case Study of "Concepts in Flux"', *Studies in History and Philosophy of Science* **21**, 563–88.

Belot, G. & J. Earman (1997). 'Chaos Out of Order: Quantum Mechanics, the Correspondence Principle, and Chaos', *Studies in History and Philosophy of Modern Physics* **28B**, 147–82.

Belousek, D. (1998). 'Ontological Commitments and Theory Appraisal in the Interpretation of Quantum Mechanics' (unpublished PhD dissertation, University of Notre Dame).

Berndl, K., M. Daumer, D. Dürr, S. Goldstein & N. Zanghì (1995). 'A Survey of Bohmian Mechanics', *Il Nuovo Cimento* **110B**, 737–50.

Berry, M. V. (1987). 'The Bakerian Lecture. Quantum Chaology' in Berry et al. (eds.) (1987), pp. 183–98.

(1989). 'Quantum Chaology, Not Quantum Chaos', *Physica Scripta* **40**, 325–6.

Berry, M. V., I. C. Percival & N. O. Weiss (eds.) (1987). *Dynamical Chaos* (Princeton, NJ: Princeton University Press).

Blaut, A. & J. K. Glikman (1996). 'Quantum Potential Approach to a Class of Quantum Cosmological Models', *Classical and Quantum Gravity* **13**, 39–49.

Bohm, D. (1952a). 'A Suggested Interpretation of Quantum Theory in Terms of "Hidden" Variables, I and II', *Physical Review* **87**, 166–93.

(1952b). 'Reply to Criticism of a Causal Re-Interpretation of the Quantum Theory', *Physical Review* **87**, 389–90.

Brown, H. R. & R. Harré (eds.) (1990). *Philosophical Foundations of Quantum Field Theory* (Oxford: Clarendon Press).

Brun, T. A. (1995). 'An Example of the Decoherence Approach to Quantum Dissipative Chaos', *Physics Letters A* **206**, 167–76.

Callender, C. & R. Weingard (1994). 'The Bohmian Model of Quantum Cosmology,' in Hull et al. (eds.) (1994), pp. 218–27.

Casati, G., B. V. Chirikov, F. M. Izrailev & J. Ford (1979). 'Stochastic Behavior of a Quantum Pendulum Under a Periodic Perturbation' in Casati and Ford (eds.) (1979), pp. 334–52.

Casati, G. (ed.) (1985). *Chaotic Behavior in Quantum Systems* (New York: Plenum Publishing Co.).

Casati, G. & J. Ford (eds.) (1979). *Stochastic Behavior in Classical and Quantum Hamiltonian Systems* (*Lecture Notes in Physics 93*) (New York: Springer-Verlag).

Caves, C. M. & R. Schack (1997). 'Unpredictability, Information, and Chaos: Pursuing Alternatives', *Complexity* (USA) **3** (1), 46–60.

Challinor, A., A. Lasenby, S. Somaroo, C. Doran & S. Gull (1997). 'Tunneling Times of Electrons', *Physics Letters A* **227**, 143–52.

Chirikov, B. V. (1991). 'Time-Dependent Quantum Systems' in Giannoni et al. (eds.) (1991), pp. 443–545.

Clifton, R. & M. Dickson (1998). 'Locality and Lorentz-Covariance in the Modal Interpretation of Quantum Mechanics', in Dieks and Vermaas (eds.) (1998).

Clifton, R. (ed.) (1996). *Perspectives on Quantum Reality* (Dordrecht: Kluwer).

Cohen, R. S. & J. Stachel (eds.) (1997). *Experimental Metaphysics* (Dordrecht: Kluwer).

Cushing, J. T. (1994). *Quantum Mechanics: Historical Contingency and the Copenhagen Hegemony* (Chicago, IL: University of Chicago Press).

(1995) 'Quantum Tunneling Times: A Crucial Test for the Causal Program?', *Foundations of Physics* **25**, 269–80.

(1996). 'What Measurement Problem?' in Clifton (ed.) (1996), pp. 167–81.

(1997). 'It is the Theory Which Decides What We Can Observe,' in Cohen & Stachel (eds.) (1997), pp. 13–23.

Cushing, J. T., A. Fine and S. Goldstein (eds.) (1996). *Bohmian Mechanics and Quantum Theory: An Appraisal* (Dordrecht: Kluwer).

Cushing, J. T. and E. McMullin (eds.) (1989). *Philosophical Consequences of Quantum Theory: Reflections on Bell's Theorem* (Notre Dame, IN: University of Notre Dame Press).

Davies, P. (ed.) (1989). *The New Physics* (Cambridge: Cambridge University Press).

Deutsch, D. (1997). *The Fabric of Reality* (London: Allen Lane).

Dewdney, C., L. Hardy & E. J. Squires (1993). 'How Late Measurements of Quantum Trajectories Can Fool a Detector', *Physics Letters A* **184**, 6–11.

Dewdney, C. & Z. Malik (1996). 'Measurement, Decoherence and Chaos in Quantum Pinball', *Physics Letters A* **229**, 183–8.

Dieks, D. and P. Vermaas (eds.) (1998). *The Modal Interpretation of Quantum Mechanics* (Dordrecht: Kluwer).

Dürr, D., W. Fusseder, S. Goldstein & N. Zanghì (1993). 'Comment on "Surrealistic Bohm Trajectories"', *Zeitschrift für Naturforschung* **48a**, 1261–2.

Dürr, D., S. Goldstein & N. Zanghì (1992a). 'Quantum Equilibrium and the Origin of Absolute Uncertainty', *Journal of Statistical Physics* **67**, 843–907.

(1992b). 'Quantum Chaos, Classical Randomness, and Bohmian Mechanics', *Journal of Statistical Physics* **68**, 259–70.

Englert, B.-G., M. O. Scully, G. Süssmann and H. Walther (1992). 'Surrealistic Bohm Trajectories', *Zeitschrift für Naturforschung* **47a**, 1175–86.

(1993). 'Reply to Comment on "Surrealistic Bohm Trajectories"', *Zeitschrift für Naturforschung* **48a**, 1263–4.

Faisal, F. H. M. & U. Schwengelbeck (1995). 'Unified Theory of Lyapunov Exponents and a Positive Example of Deterministic Quantum Chaos', *Physics Letters A* **207**, 31–6.

Ford, J. (1988). 'Quantum Chaos, Is There Any?' in Bai-Lin (ed.) (1988), pp. 128–47.

(1989). 'What Is Chaos That We Should Be Mindful of It?' in Davies (ed.) (1989), pp. 348–72.

Ford, J. & G. Mantica (1992). 'Does Quantum Mechanics Obey the Correspondence Principle? Is It Complete?', *American Journal of Physics* **60**, 1086–98.

French, S. & M. L. G. Redhead (1988). 'Quantum Physics and the Identity of Indiscernibles', *British Journal for the Philosophy of Science* **39**, 233–46.

Frisk, H. (1997). 'Properties of the Trajectories in Bohmian Mechanics', *Physics Letters A* **227**, 139–42.

Giannoni, M.-J., A. Voros and J. Zinn-Justin (eds.) (1991). *Chaos and Quantum Physics* (*Les Houches, Session 52*) (Amsterdam: North-Holland Publishing Co.).

Goldstein, H. (1950). *Classical Mechanics* (Reading, MA: Addison-Wesley Publishing Co.).

Guckenheimer, J. & P. Homes (1983). *Nonlinear Oscillations, Dynamical Systems, and Bifurcations of Vector Fields* (New York: Springer-Verlag).

Guglielmo, I. & M. Pettini (1996). 'Regular and Chaotic Quantum Motions', *Physics Letters A* **212**, 29–38.

Güntherodt, H.-J. & R. Wiesendanger (eds.) (1993). *Scanning Tunneling Microscopy III* (Berlin: Springer-Verlag).

Gutzwiller, M. C. (1990). *Chaos in Classical and Quantum Mechanics* (New York: Springer-Verlag).

Haake, F. (1991). *Quantum Signatures of Chaos* (Berlin: Springer-Verlag).

Hirsch, M. W. & Smale (1974). *Differential Equations, Dynamical Systems, and Linear Algebra* (New York: Academic Press).

Holland, P. R. (1993). *The Quantum Theory of Motion: An Account of the de Broglie–Bohm Causal Interpretation of Quantum Mechanics* (Cambridge: Cambridge University Press).

(1996). 'Is Quantum Mechanics Universal?' in Cushing et al. (1996), pp. 99–110.

Horiguchi, T. (1994). 'Quantum Potential Interpretation of the Wheeler–De Witt Equation', *Modern Physics Letters A* **9**, 1429–43.

Hull, D., M. Forbes & R. M. Burian (eds.) (1994). *PSA 1994*, Vol. 1 (East Lansing, MI: Philosophy of Science Association).

Hull, D., M. Forbes & K. Okruhlik (eds.) (1993). *PSA 1992*, Vol. 2 (East Lansing, MI: Philosophy of Science Association).

Iacomelli, G. & M. Pettini (1996). 'Regular and Chaotic Quantum Motions', *Physics Letters A* **212**, 29–38.

Izrailev, F. M. (1990) 'Simple Models of Quantum Chaos: Spectrum and Eigenfunctions', *Physics Reports* **196**, 299–392.

Jarrett, J. P. (1984). 'On the Physical Significance of the Locality Conditions in the Bell Arguments', *Nous* **18**, 569–89.

Leavens, C. R. & G. C. Aers (1993). 'Bohm Trajectories and the Tunneling Time Problem', in Güntherodt and Wiesendanger (eds.) (1993), pp. 105–40.

Lichtenberg, A. J. & M. A. Lieberman (1992). *Regular and Chaotic Dynamics* (2nd edn, New York: Springer-Verlag).

Malik, Z. & C. Dewdney (1994). 'Quantum Mechanics, Chaos and the Bohm Theory', University of Portsmouth preprint.

Maudlin, T. (1994). *Quantum Non-Locality and Relativity* (Oxford: Blackwell).

McCauley, J. L. (1993). *Chaos, Dynamics, and Fractals* (Cambridge: Cambridge University Press).

Nakamura, K. (1993). *Quantum Chaos* (Cambridge: Cambridge University Press).

Parmenter, R. H. & R. W. Valentine (1995). 'Deterministic Chaos and the Causal Interpretation of Quantum Mechanics', *Physics Letters A* **201**, 1–8. [*Erratum* **213** (1996), 310.]

(1997). 'Chaotic Causal Trajectories Associated With a Single Stationary State of a System of Noninteracting Particles', *Physics Letters A* **227**, 5–14.

Rasband, S. N. (1990). *Chaotic Dynamics of Nonlinear Systems* (New York: John Wiley & Sons).

Redhead, M. L. G. (1983). 'Quantum Field Theory for Philosophers' in Asquith & Nickles (eds.) (1983), pp. 57–99.

(1987). *Incompleteness, Nonlocality, and Realism* (Oxford: Clarendon Press).

(1990). 'A Philosopher Looks at Quantum Field Theory' in Brown & Harré (eds.) (1990), pp. 9–23.

(1995). *From Physics to Metaphysics* (Cambridge: Cambridge University Press).

Redhead, M. L. G. & P. Teller (1991). 'Particles, Particle Labels, and Quanta: The Toll of Unacknowledged Metaphysics', *Foundations of Physics* **21**, 43–62.

(1992). 'Particle Labels and the Theory of Indistinguishable Particles in Quantum Mechanics', *British Journal for the Philosophy of Science* **43**, 201–18.

Rohrlich, F. (1988). 'Pluralistic Ontology and Theory Reduction in the Physical Sciences', *British Journal for the Philosophy of Science* **39**, 295–312.

Schack, R. & C. M. Caves (1993). 'Hypersensitivity to Perturbations in the Quantum Baker's Map', *Physical Review Letters* **71**, 525–8.

Schack, R., G. M. D'Ariano & C. M. Caves (1994). 'Hypersensitivity to Perturbation in the Quantum Kicked Top', *Physical Review E* **50**, 972–87.

Schwengelbeck, U. & F. H. M. Faisal (1995). 'Definition of Lyapunov Exponents and KS Entropy in Quantum Dynamics', *Physics Letters A* **199**, 281–6.

Selleri, F. (1994). 'Theories Equivalent to Special Relativity', in Barone & Selleri (eds.) (1994), pp. 181–92.

Sengupta, S. & P. K. Chattaraj (1996). 'The Quantum Theory of Motion and Signatures of Chaos in the Quantum Behaviour of a Classically Chaotic System', *Physics Letters A* **215**, 119–27.

Shtanov, Yu. V. (1996). 'Pilot-Wave Quantum Cosmology', *Physical Review D* **54**, 2564–70.

Spiller, T. P. & J. F. Ralph (1994). 'The Emergence of Chaos in an Open Quantum System', *Physics Letters A* **194**, 235–40.

Tabor, M. (1989). *Chaos and Integrability in Nonlinear Dynamics: An Introduction* (New York: John Wiley & Sons).

Valentini, A. (1991a). 'Signal-Locality, Uncertainty, and the Subquantum H-Theorem. I', *Physics Letters A* **156**, 5–11.

(1991b). 'Signal-Locality, Uncertainty, and the Subquantum H-Theorem. II', *Physics Letters A* **158**, 1–8.

(1992). 'On the Pilot-Wave Theory of Classical, Quantum and Subquantum Physics', PhD dissertation, International School for Advanced Studies, Trieste, Italy.

(1996). 'Pilot-Wave Theory of Fields, Gravitation and Cosmology' in Cushing et al. (1996), pp. 45–66.

(1997). 'On Galilean and Lorentz Invariance in Pilot-Wave Dynamics', *Physics Letters A* **228**, 215–22.

(1999), *Pilot-Wave Theory: An Alternative Approach to Modern Physics* (Berlin: Springer-Verlag). (forthcoming)

Vink, J. C. (1992). 'Quantum Potential Interpretation of the Wave Function of the Universe', *Nuclear Physics B* **369**, 707–28.

Wilkinson, P. B., T. M. Fromhold, L. Eaves, F. W. Sheard, N. Miurs & T. Takamasu (1996). 'Observation of "Scarred" Wavefunctions in a Quantum Well with Chaotic Electron Dynamics', *Nature* **380**, 608–10.

Zurek, W. H. (1991). 'Decoherence and the Transition from Quantum to Classical', *Physics Today* **44** (10), 36–44.

(1993). 'Preferred States, Predictability, Classicality and the Environment-Induced Decoherence', *Progress of Theoretical Physics* **89**, 281–312.

6

Strange positions

G. N. FLEMING AND J. BUTTERFIELD

1. Introduction: localization and Lorentz-invariant Quantum Theory

The current status of localization and related concepts, especially localized statevectors and position operators, within Lorentz-invariant Quantum Theory (LIQT) is ambiguous and controversial.[1] Ever since the early work of Newton & Wigner (1949), and the subsequent extensions of their work, particularly by Hegerfeldt (1974, 1985), it has seemed impossible to identify localized statevectors or position operators in LIQT that were not counterintuitive – strange – in one way or another; the most striking strange property being the superluminal propagation of the localized states.

The ambiguous and controversial status of these concepts arises from the varied reactions that workers have to the strange properties. Some regard them, particularly the superluminal propagation, as utterly unacceptable, and conclude that no precise concepts of localized statevectors and position operators exist in LIQT (Wigner 1973, pp. 325–7; 1983, pp. 310–13; Malament 1996). Others downplay the whole issue, on the grounds that current theoretical and experimental practice has no need of sharply formulated concepts of localization, localized states etc. (Birrell & Davies 1984, pp. 48–59; Haag 1992, p. 34).

Still others, including ourselves, believe that the superluminal propagation does not lead to causal contradictions, and is not in conflict with

It is a pleasure to dedicate this essay to Michael Redhead, especially in view of our discussions of the topic over the years. We thank many friends and seminar audiences, in the US and Europe, for discussions; and we especially thank Paul Busch, Klaas Landsman and David Malament and Simon Saunders for comments on this paper. Work on the essay began in 1993, during a visit to Prof Ghirardi at ICTP and the University of Trieste. It is a pleasure to thank him, ICTP and SISSA, for very generous hospitality and financial support. JB also thanks the Mrs L.D. Rope Third Charitable Settlement for supporting sabbatical leave.

[1] We give a glossary of acronyms at the end of the paper.

available empirical data: so that it should not be ruled out. We also believe that the other strange properties are merely unfamiliar novelties of LIQT which we must simply learn to accept. (Like the superluminal propagation, they do not lead to causal contradictions, nor conflict with available data.) So in this essay, we aim to help remove the ambiguous and controversial status of localization concepts in LIQT.

Overall, our strategy will be to assess a number of localization concepts, and thereby clarify the often complex relationships among them. This overall strategy breaks down into two parts, corresponding to sections 2 to 4 and sections 5 to 12.

In part 1, we will summarize some well-established material about localization. First, we will report Newton & Wigner's approach to localization, and the two strange properties which their solutions were found to possess: superluminal propagation and delocalization under a passive Lorentz boost (section 2). Here we will emphasize the similarities with the much simpler and non-controversial case of non-relativistic, i.e. Galilean-invariant Quantum Theory (GIQT). Although no conceptual problems about localization occur in GIQT, considering this case will make it easy to anticipate, on the basis of familiarity with classical special relativity, where subtleties in LIQT will arise.

Then we turn back to work done before Newton & Wigner's: we summarize the various difficulties in defining for the Klein–Gordon and Dirac equations 'well-behaved' position eigenstates. In section 3, we treat the Klein–Gordon case; and in section 4, the Dirac case. Once we are confronted with (exhausted by!) all these difficulties, we are prompted to ask: is there any standpoint from which the Newton–Wigner states, despite their strange properties under time-evolution and passive Lorentz boost, make physical sense?

Part 2 (sections 5 to 12) answers 'Yes' to this question. After a prospectus (section 5) and discussion of the idea of an 'event' operator (section 6), we turn to *classical* Lorentz-covariant position variables – and first describe three different parametrizations of them (section 7). These are equivalent for a point-particle; but for a localizable property of an extended system (such as the centre of energy or of charge), only one parametrization, the hyperplane-dependent parametrization, is convenient (section 8). Of the three, it is also the only one that can be consistently quantized; where the quantization procedure is the usual requirement that the Poincaré transformation equation for the classical position variable should serve as the transformation equation for the expectation value of the quantum position operator (section 9). This in effect establishes the Yes answer to the above question; though admittedly, the strange properties of localized states under time-evolution and passive Lorentz boost are not *wholly* unsurprising from the standpoint that part 2 develops. The closing three sections discuss in order: examples (section 10); the strange properties (section 11); and some recent criticisms of our standpoint, by Malament and by Saunders (section 12).

We close this introduction with six preliminary remarks.

(1) We will use 'LIQT' and 'GIQT' to refer to what may be called 'framework theories' of which specific quantum theories are examples. By a 'specific quantum theory' we intuitively mean a complete quantum theory of a specific physical system. In mathematical terms, we expect such a theory to identify a complete set of basic dynamical variables sufficient to determine, in principle, the commutators of any two operators in the theory; and to specify the functional dependence of the generators of the spacetime symmetry group on the basic dynamical variables. On the other hand, LIQT and GIQT need only postulate: (i) the existence of a preferred unitary or projective-unitary representation of the spacetime symmetry group (Poincaré or Galilean group, respectively) on the state space; (ii) the identification of the generators of these group representations with physical properties of the system in the usual way: viz. the spatial translation generators with total momentum, the rotation generators with total angular momentum, the time translation generator with total energy, and similarly for the boost generators (cf. eqns 111–113 for their details). Thus we do not presume a quantum field theory by LIQT, but we certainly allow for one; similarly, we allow for one in GIQT, provided the mass superselection rule is satisfied. (Of course, at various points in the paper we will consider specific quantum theories; and we will there specify the additional structure.) This leads to our second preliminary remark.

(2) We want to dispel once and for all a widespread misconception about our topic, viz. that the whole issue pertains exclusively to single particle states, or at most, to systems with a fixed number of particles. This misconception is widespread particularly among those who dismiss the localization problem as unimportant. For their reason is usually that, since the physically important LIQT's are theories of interacting quantized fields (local fields, one might add) in which particle numbers are not only changing with time but are also usually indefinite, the question of the properties of proposed localized statevectors and position operators for individual particles is just not important.

In reply, we stress that the localization problems we address here are in no way confined to individual particles or systems of a fixed number of particles. While sections 3 and 4 will focus on such systems, our later discussion will include such topics as the location of the centre of the energy distribution for a system of interacting quantized fields (section 10). Furthermore, we will find that all the strangenesses of localization of particles are visited undiminished on fields – and can also be understood and dealt with in the same ways.

The novelty of the field theoretic case is that local quantum field theory comes already equipped with spatio-temporal concepts that are, however, only indirectly and non-locally related to the localization concepts closely associated with localized statevectors and position operators. The examination of the relationship between these two kinds of localization concepts is an important topic which we address briefly in section 12: for discussion, cf. Fulling (1989, pp. 71–3) and Fleming (1996, pp. 19–26). This brings us to our third preliminary remark.

(3) We emphasize that the idea of *different* localization concepts is not mysterious. For complex quantum systems (i.e. those with many degrees of freedom, i.e. many independent dynamical variables), it is not surprising that there are many associated concepts of localized states and position operators; (quite apart from the localized states and position operators for the various *subsystems* of the complex system). For example, in GIQT we have the centre of mass and, if the system has non-vanishing total electric charge, the centre of charge; and if, for example, the system consists of electrons and protons, the two centres will be different. Furthermore, the centre of kinetic energy, and the centres of the nth powers of mass or charge or kinetic energy can be identified, albeit to perhaps little purpose for physics.

We mention these varieties here because we will later (sections 10, 11) argue for the legitimacy and relevance, in LIQT, of several distinct concepts of localized statevector and position operator, even for *single particle systems*; which undermines any tendency to think of such systems as *point* particles; (Kalnay & Torres 1973; Lorente & Roman 1974). This leads to our fourth remark.

(4) Another approach to localization in LIQT, leading to a form of hyperplane dependence closely related to that described in our part 2 (especially sections 9 *et seq.*), has been developed by workers studying the problems of quantizing classical systems on phase space; (cf. Ali 1998; Landsman 1999; where further references are given). This approach is more rigorous and

systematic than ours; but ours, by being more physically intuitive and informal, is also more inclusive. In particular, by focusing on coordinates that can play a fundamental role in a phase space formulation, the phase space quantization procedure restricts itself, for massive systems at least, to yielding position operators with commuting components, essentially the Newton–Wigner operator for particles. Our broader, if looser, net will encompass collective coordinates of extended systems and will lead quite naturally to the field theoretic centre of energy position operator, which has non-commuting components; (and from which the Newton–Wigner operator will be algebraically constructable, cf. eqn 117).

(5) We should briefly state the position we adopt here about the conditions under which physical quantities have values. In short, we will adopt the orthodox 'eigenvalue–eigenstate' link: a quantum system possesses a definite value for a quantity just in case its statevector is an eigenvector of the operator representing the quantity (and the eigenvalue is the value in question). Thus for us the possession of a value for a quantity is in no way tied to the execution of a measurement. On the other hand, we accept that a notion of measurement is crucial, to enable us to give a probability interpretation of inner products of statevectors and eigenvectors. (As we will see in section 2, the Newton–Wigner orthogonality requirement is motivated by the probability interpretation.)

We of course recognize that the eigenvalue–eigenstate link is very controversial. But we adopt it here, not because we endorse it – though we admit to favouring the related doctrine that state-reduction is a real physical process. Rather, we believe that the controversy about it does not affect the issues we will discuss.

(6) Finally, since we will often have to compare the description of quantum states and quantities from the standpoints of different reference frames, we should stress that in quantum theory the statevectors do *not* represent the intrinsic physical situation (in the Heisenberg picture: the intrinsic history) of the system. Rather they represent the relationships of the physical situation or history to the reference frame for which they are employed. This is why a change of reference frame entails a change of statevector for the same physical history (a passive transformation).

Similarly the operators representing quantities in quantum theory do not represent directly the invariant quantity, but rather the relationship of (the range of possible values of) the quantity to the reference frame. Consequently, physical quantities are usually defined in a frame-dependent way (e.g. in the Heisenberg picture: the Cartesian components of momentum at a time $t, p_k(t)$); though the operator representing the quantity may not change at all under a change of reference frame (in the example this will be so unless the time coordinate, t, is reassigned by the transformation). Accordingly, combining with (5): the eigenvectors of a quantity represent the relationship to the reference frame of physical situations or histories in which the (usually frame-dependent) quantity possesses the corresponding eigenvalue.

2. The Newton–Wigner analysis: going orthogonal under the Euclidean group

We now describe Newton & Wigner's (1949: hereafter NW) approach to the task of identifying localized statevectors. As mentioned in section 1, their leading idea is one that applies equally to GIQT and LIQT. For they characterize these statevectors in terms of the way they change under spatial translations and rotations, i.e. under the Euclidean group. But the Euclidean group is a subgroup of both the Galilean group (the spacetime symmetry group of GIQT), and the Poincaré group (the spacetime symmetry group of LIQT). In other words, what NW require of localized statevectors makes no reference to dynamical evolution or the relationship between relatively moving frames of reference.

NW then showed that for what they called 'elementary systems' (see below), the localized statevectors are uniquely determined by their Euclidean group properties. So, in a sense, one is dealing in both GIQT and LIQT with the same set of localized statevectors; and one simply has to investigate the different ways those statevectors change under the non-Euclidean elements of the Galilean and Poincaré groups.

NW's requirements on localized statevectors stem from the following point:

> If Y is a state localized in a volume V at a definite time; and Ψ' is a state localized in a volume V' at the same time; and V and V' are disjoint: then the statevectors corresponding to these states in a given reference frame F, $|\Psi >$ and $|\Psi' >$ respectively, must be orthogonal.

For if the state Ψ deserves to be called 'localized', then measuring whether the system in the state Ψ is in the volume V' should be guaranteed to yield a negative result; i.e. the probability of a positive result should be zero. But similarly, if Ψ' deserves to be called 'localized' in V', then its statevector

$|\Psi'>$ should lie in the subspace corresponding to that positive result. So it follows by the Born rule that $|\Psi>$ and $|\Psi'>$ must be orthogonal.

Now consider the result of applying a spatial translation, with displacement vector $\mathbf{a}$, to the state Ψ (an active transformation). This produces a state localized in the similarly displaced volume $V_{\mathbf{a}}$ at the same instant of time. But for bounded V, $V_{\mathbf{a}}$ will be disjoint from V for sufficiently large $\mathbf{a}$. Consequently $|\Psi>$ and $|\Psi_{\mathbf{a}}>$ will be orthogonal for all sufficiently large $\mathbf{a}$. Similarly if V is far enough away from frame F's spatial origin of coordinates, then the application of a rotation R, about that origin, to Ψ, yielding a state localized in the similarly rotated volume V_R, will, if R is large enough, make V_R disjoint from V, and therefore $|\Psi_R>$ orthogonal to $|\Psi>$.

Thus localized states correspond to statevectors that can be rendered orthogonal to themselves by application of the unitary operators representing elements of the Euclidean group. The smaller the volume V, the larger the set of unitary operators that yield a statevector orthogonal to the original – thereby constraining the statevector.

In fact, NW did not consider partially localized states as we have just done. (The first rigorous treatment of such states was Wightman 1962.) They dealt directly with the mathematical limit of point-localized vectors, requiring that they are made orthogonal to themselves by *all* spatial translations. They also required that any point-localized state subject to a rotation about that point remained localized at that point; and that translation of any state so localized rendered it orthogonal to all the localized states obtained by rotating about the given point.

It is intuitively clear that these requirements cannot alone determine what are the localized statevectors. For the system could have various kinds of internal structure, which are not constrained by these requirements. So in order to uniquely identify the localized statevectors, NW concerned themselves with what are called (following Wigner 1939) 'elementary systems'. Intuitively, these are systems without internal excitations or deviations from a specified internal structure. This means that all their states are to be obtainable by superposing the images, under elements of the spacetime symmetry group, of any state. Accordingly, they are technically defined as systems whose state spaces support an irreducible representation of the spacetime symmetry group.

Elementary systems, thus defined, include what we call 'elementary particles'. But they also include systems we regard as highly structured, but restricted to subspaces of their normally available states. As NW point out (1949, p. 400) the Hydrogen atom in its ground state is an elementary system; and, we might add, so is the Helium atom and the Lead nucleus in their ground states; and, were it not for decay back towards the ground state, the subspaces of states corresponding to a given excitation of a complex system.

NW showed that with their restriction to massive elementary systems, their orthogonality requirements were enough to uniquely identify the localized statevectors. That is: for an elementary system with spin quantum number s, the NW point-localized states, for localization at a given point, are uniquely determined to span a $2s+1$ dimensional subspace corresponding to the different possible 'spin states' for that localization. The states corresponding to localization within a volume V, that we discussed above, are then identified with linear superpositions of these NW position eigenvectors for points lying within V.

We should add that the NW analysis also yields a complete characterization of the relationship between the localized states (position eigenvectors), and the momentum eigenvectors: a characterization which is invariant under the transition from GIQT to LIQT. In this transition, it is rather the relationship between position and *velocity* eigenvectors that changes (since the velocity operator is defined in terms of the *time* rate of change of the position operator). Even here, however, the change is trivial for free elementary systems for which the momentum and velocity eigenvectors are identical. Only the functional relationship between the eigenvalues changes. (We will further discuss relativistic velocity operators and eigenvectors in sections 3 and 4.)

We emphasize that all of these statements refer to localization at a definite time. Thus, if the time is changed then, in the Heisenberg picture, the set of states localized in specified finite volumes or at points of space changes.

So far, everything seems fine: the whole analysis, with all its details, fits nicely into the standard framework of general quantum theory. But we now need to consider the results of applying, to these localized statevectors, the unitary operators corresponding to time translations and transformations to relatively moving reference frames. We must identify and interpret the transformed statevectors; and we must examine their inner products with the original statevectors.

In GIQT everything works out as one's intuition would lead one to expect. Thus if $|\mathbf{x}, \mu; t>$ is a Heisenberg picture position eigenvector at the time t with position eigenvalue $\mathbf{x}$, and μ takes on the $2s+1$ values denoting mutually orthogonal vectors spanning the system's spin state space at each $\mathbf{x}$ and t, then the result of applying a time translation operator with translation parameter b is the position eigenvector $|\mathbf{x}, \mu; t+b>$ for localization at the time $t+b$. The inner products for $b \neq 0$,

$$< \mathbf{x}, \mu; t|\mathbf{x}', \mu'; t+b >$$

are almost never zero, no matter how large the separation $|\mathbf{x} - \mathbf{x}'|$ between $\mathbf{x}$ and $\mathbf{x}'$. This is commonly said to correspond to the possibility of motion with arbitrarily high velocity; and this is what one would expect by applying the Heisenberg uncertainty principle for position and momentum/velocity to the limiting case of position eigenvectors. (For some discussion of the uncertainty principle in the relativistic domain, cf. sections 6 and 9.)

Furthermore, applying the operators corresponding to the transformation to a relatively moving reference frame yields, as one might expect, a position eigenvector in which the position eigenvalue changes with time. If the original reference frame has velocity $\mathbf{v}$ relative to the new reference frame, then the operation, applied to $|\mathbf{x}, \mu; t>$ yields, up to a phase factor depending on $\mathbf{x}$, $\mathbf{v}$, m and t, the localized statevector $|\mathbf{x} + \mathbf{v}t, \mu; t>$, in accordance with the Galilean boost

$$\mathbf{x}' = \mathbf{x} + \mathbf{v}t.$$

But in LIQT there is trouble! We will spell out more details later (especially section 11), but in short:

(i) The result of applying time translation operators to the LIQT position eigenvectors is too much *like* the result in GIQT. That is: $|\mathbf{x}, \mu; t>$ still goes to $|\mathbf{x}, \mu; t+b>$ and the inner product $< \mathbf{x}, \mu; t|\mathbf{x}', \mu'; t+b >$ *still fails to vanish for arbitrarily large* $|\mathbf{x} - \mathbf{x}'|$, including spacelike intervals between $(\mathbf{x}, t)$ and $(\mathbf{x}', t+b)$. (Cf. Wightman & Schweber (1955, esp. p. 826); Fleming (1965b, p. 963, note 5); and Hegerfeldt (1974).)

(ii) The result of applying Lorentz boost operators is too much *unlike* the result in GIQT. That is: the application of a Lorentz boost operator to $|\mathbf{x}, \mu; t>$ does not yield any localized statevector at all; i.e. *it yields a statevector that is not localized at any definite time, anywhere*, not even within a finite volume, let alone at a point.

So we have (i) superluminal propagation from localized states and (ii) complete delocalization by an operation representing a passive transformation to a relatively moving reference frame. It is of course these results of NW's analysis that have made many workers uncomfortable.

3. Relativistic wavefunctions: the Klein–Gordon case

Historically it was the obstacles to a straightforward interpretation of the Klein–Gordon (KG) and Dirac wavefunctions as position probability amplitudes that drove workers such as NW to attempt to secure a precise localization concept in LIQT. But unless one is expert in the interpretative problems of such wavefunctions, one is likely, on learning of the difficulties revealed at the end of section 2, to wonder just why the KG and Dirac wavefunctions cannot be naively interpreted as position probability amplitudes after all. In this section and the next, we will briefly review some of the reasons for this.

Our discussion is partly motivated by the fact that most textbooks do not discuss these matters. More importantly, many textbooks' treatments of the KG and Dirac equations foster (no doubt, often unwittingly) the misconception we tried to dispel in remark (2) of section 1: viz. that interpretative difficulties about localization relate only to single particle, or at least fixed particle-number, states. For as we shall see, these difficulties are related to the existence of negative energy solutions of the wave equation, and *their* interpretative problems. The main such problem is the so-called 'stability problem'. It turns out that the energy eigenvalues are unbounded from below, i.e. extend to $-\infty$, so that an arbitrarily large amount of energy can apparently be extracted from the system: one would just need an interaction to allow the system to fall down through the energy spectrum. This suggests that the equation, as a theory of a free particle, is unrealistic, not only in the trivial sense that there always are interactions, but also in that it cannot be extended to interactions.

Accordingly, the books see the problems of negative energy as 'pointing the way' to quantum field theory, where the problems are resolved by reinterpreting the solutions in terms of positive energy, oppositely-charged antiparticles. (This reinterpretation was first suggested by Pauli & Weisskopf (1934).) The idea is to use only the original positive energy states, now regarded as positive *frequency* states, for the particles, and only the negative *frequency* states for the antiparticles; energy now being the product of Planck's constant and the absolute value of the frequency.

So we want to emphasize that although the difficulties about localization are related to the negative energy problems, they nevertheless *persist* after we make the usual restriction to positive energies – whether for a single particle, or fixed particle-number, or for fields. Indeed, our main point, in both this section and the next, will be that once we make this restriction, we must confront the NW concept of position, and thereby the strange properties, (i) and (ii), at the end of section 2.

The situation is similar as regards the other main interpretative problem besetting the KG equation: the non-positivity of the natural density (i.e. time-component of the natural conserved 4-current density) defined by the KG wavefunction. That is to say: we shall see below that in the KG case, the interpretative difficulties about localization are related to this density's not being positive definite. And many books say: (a) this non-positivity prompts the density being taken as a charge, not probability, density (once we multiply it by the electronic charge); and (b) this motivates both: a single charge, but not single particle, interpretation of the KG equation; and also the Dirac equation – whose natural density is positive definite, and so could be taken as a probability density.

As it happens, we have our doubts about (a) and (b) : e.g. (a) does not apply to the case where the KG formalism is used to describe neutral spinless particles. But our main point is that, even if (a) and (b) are fair enough, they foster the misconception that the difficulties about localization somehow disappear for the Dirac equation, or at least once we make the transition to field theory. Again we emphasize: they don't!

Here are some examples of treatments (in otherwise excellent books!) which foster these misconceptions. First, for the KG case: the non-positivity of the density is said to prompt a charge interpretation; and this non-positivity, and the stability problem about negative energy solutions, are said to prompt field theory, by: Itzykson & Zuber (1985, p. 48), Messiah (1961, p. 888), Ryder (1985, p. 31) and Schiff (1968, p. 468). Secondly, for the Dirac case: the stability problem, and the difficulties besetting Dirac's ingenious proposed solution to it (a 'sea', filling all the negative energy states), are said to prompt field theory, by: Itzykson & Zuber (1985, pp. 84–5), Messiah (1961, pp. 953–6), Ryder (1985, pp. 47–8) and Schiff (1968, p. 488).

Of course, there are exceptions. We especially note the books of Schweber and Greiner. Schweber's masterly (1961) discusses localization for both the KG case (pp. 60–2) and the Dirac case (pp. 94–5). More precisely, he describes the NW position operator for these two cases (and its eigenvectors for the KG case); and for the Dirac case relates it to the Foldy–Wouthuysen representation. He also maintains, reasonably enough, that although the stability problem shows that the theory of a single free particle is very limited, it does not amount to an *internal* defect of the theory: for such a particle by definition does not undergo interactions (pp. 56, 96, 99). We will return to this viewpoint below (after eqn 13).

Greiner devotes a volume (1997) of his monumental series in physics, to relativistic quantum mechanics (i.e. without field quantization). Like Schweber, he describes the position operator for both the KG case (pp. 73–4) and the Dirac case (pp. 279, 282), and eigenvectors for the former (pp. 75, 78–83); but he gives much more detail, in particular describing the KG-analogue of the Foldy–Wouthuysen representation, viz. the Feshbach–Villars representation.

Accordingly, in our brief review of some of the difficulties about localization, in this section and the next, we have selected material that is both elementary, and supplements (partially!) Schweber's & Greiner's material. This means that in this section we will search in a naive way for position eigenfunctions for the KG case; but we shall not take the space to 'succeed', i.e. to exhibit the NW operator, nor to introduce the Feshbach–Villars representation. (The next section will have to have similar Draconian limitations.)

So much by way of introduction. We turn to the KG equation for the free evolution of a single spinless quantum particle of rest mass m:

$$(\Box + \kappa^2)\phi(x) = 0, \tag{1}$$

where $\kappa := mc/\hbar$. The solution, $\phi(x)$, is taken to be a single complex function transforming under Poincaré transformations as a local scalar field;

$$\phi'(x') = \phi(x). \tag{2}$$

Any physical interpretation of ϕ in terms of possible positions of the particle must respect the secure interpretation of the Fourier transform of ϕ as the amplitude for the momentum probability density. The general solution of eqn 1 has the Fourier representation,

$$\begin{aligned}\phi(x) = \phi(\mathbf{x}, x^0) = (2\pi\hbar)^{-3/2} \int (\mathrm{d}^3 p/2p^0)[\tilde{\phi}_+(\mathbf{p}) \exp[(i/\hbar)(\mathbf{px} - p^0 x^0)] \\ + \tilde{\phi}_-(\mathbf{p}) \exp[(i/\hbar)(\mathbf{px} + p^0 x^0)]]\end{aligned} \tag{3}$$

where $p^0 = |\sqrt{(m^2c^2 + \mathbf{p}^2)}|$. The momentum space volume element is chosen to be $\mathrm{d}^3p/2p^0$, since that is the Lorentz-invariant measure on the mass shell (associated with rest mass m), consisting of all 4-vectors $p = (p^0, \mathbf{p})$ such that $p^2 = (p^0)^2 - \mathbf{p}^2 = m^2$; and this choice allows the Fourier coefficients, $\tilde{\phi}_\pm(\mathbf{p})$ to be momentum space Lorentz scalar functions. These functions are called the positive energy (frequency) and negative energy (frequency) momentum amplitudes respectively. The first exponential function of $\mathbf{x}$ and x^0 is just the complex plane wave proposed by de Broglie for a state with precise momentum and (positive!) energy; and the second exponential is its negative energy analogue. Accordingly, the conditional probability densities for the possession of momentum **p** given positive or negative energy are;

$$P_{\pm}(\mathbf{p}) \equiv (|\tilde{\phi}_{\pm}(\mathbf{p})|^2/2p^0)/\int(\mathrm{d}^3p'/2p'^0)|\tilde{\phi}_{\pm}(\mathbf{p}')|^2 \tag{4}$$

respectively.

This suggests that it is appropriate to define the following positive-definite inner product:

$$<\phi_2|\phi_1>:=\int(\mathrm{d}^3p/2p^0)[\tilde{\phi}_{2+}(\mathbf{p})^*\tilde{\phi}_{1+}(\mathbf{p})+\tilde{\phi}_{2-}(\mathbf{p})^*\tilde{\phi}_{1-}(\mathbf{p})] \tag{5}$$

This inner product, however, cannot be expressed as a spatial integral with an integrand local in $\phi_2(\mathbf{x}, x^0)$ and $\phi_1(\mathbf{x}, x^0)$. The only Lorentz-invariant and time-independent inner product on the KG solution space that is so expressible is:

$$(\phi_2|\phi_1):=i\hbar\int\mathrm{d}^3x\phi_2(\mathbf{x},x^0)^*\left(\overleftrightarrow{\partial/\partial x^0}\right)\phi_1(\mathbf{x},x^0), \tag{6a}$$

and it has these desirable properties precisely because the integrand is the time-component of a locally conserved 4-vector current

$$j_{21,\mu},(x):=i\hbar\phi_2(\mathbf{x},x^0)^*\left(\overleftrightarrow{\partial/\partial x^\mu}\right)\phi_1(\mathbf{x},x^0) \tag{6b}$$

just as we would expect of a position probability density.

Unfortunately, this inner product is not positive definite. This follows from the fact that, since the KG equation is second-order in time, ϕ and $(\partial/\partial x^0)\phi$ can be fixed independently at a given time. Another way to see this is to express the inner product in terms of momentum space amplitudes;

$$(\phi_2|\phi_1):=\int(\mathrm{d}^3p/2p^0)[\tilde{\phi}_{2+}(\mathbf{p})^*\tilde{\phi}_{1+}(\mathbf{p})-\tilde{\phi}_{2-}(\mathbf{p})^*\tilde{\phi}_{1-}(\mathbf{p})]. \tag{6c}$$

Consequently, the spatial integrand in eqn 6a, for the case of $\phi_2 = \phi_1$, cannot be interpreted as a position probability density.

So neither inner product is wholly satisfactory. Comparing them we note that for wave functions lying wholly within an energy sign eigenspace (i.e. one of the two subspaces with energy of a definite algebraic sign), the two inner products are closely related. In the positive energy subspace they are identical, while in the negative energy subspace they are the negatives of one another. When expressed in terms of the positive and negative energy parts of $\phi(x)$

$$\phi_{\pm}(\mathbf{x}, x^0) := (2\pi\hbar)^{-3/2} \int (\mathrm{d}^3p/2p^0)\tilde{\phi}_{\pm}(\mathbf{p}) \exp[(i/\hbar)(\mathbf{px} \mp p^0x^0)] \tag{7}$$

the positive definite inner product has the form

$$\begin{aligned} < \phi_2|\phi_1 >:= i\hbar \int \mathrm{d}^3x[\phi_{2+}(\mathbf{x}, x^0)^* * \left(\overleftrightarrow{\partial/\partial x^0}\right)\phi_{1+}(\mathbf{x}, x^0) \\ - \phi_{2-}(\mathbf{x}, x^0)^* * \left(\overleftrightarrow{\partial/\partial x^0}\right)\phi_1(\mathbf{x}, x^0)]; \end{aligned} \tag{8}$$

but, again, the integrands, $\pm\phi_{2\pm}(\mathbf{x}, x^0)^* * (\overleftrightarrow{\partial/\partial x^0})\phi_{1\pm}(\mathbf{x}, x^0)$ are not themselves positive semidefinite forms. And this means that even if we confined ourselves to an energy sign eigenspace we could not regard such integrands as position probability densities.

In view of these difficulties one may be inclined to systematically look for the eigenfunctions of 'the' position operator. Since the KG equation is second-order in time, a unique solution is determined by ϕ and $(\partial/\partial x^0)\phi$ over all space at some time. So to define a linear operator on the KG solution space, one must specify its action on both ϕ and $(\partial/\partial x^0)\phi$ at some definite time. So suppose we define 'the' position operator $\boldsymbol{X}$ by:

$$[\boldsymbol{X}\phi](\mathbf{x}, x^0) = \mathbf{x}\phi(\mathbf{x}, x^0) \tag{9a}$$

$$[\boldsymbol{X}(\partial/\partial x^0)\phi](\mathbf{x}, x^0) = \mathbf{x}(\partial/\partial x^0)\phi(\mathbf{x}, x^0) \tag{9b}$$

The eigenfunctions, $\xi_{\mathbf{y},y_0}(\mathbf{x}, x^0)$ must then satisfy

$$(\mathbf{x} - \mathbf{y})\xi_{\mathbf{y},y_0}(\mathbf{x}, y^0) = 0 \tag{10a}$$

$$(\mathbf{x} - \mathbf{y})[(\partial/\partial x^0)\xi_{\mathbf{y},y_0}(\mathbf{x}, x^0)|_{x_0=y_0}] = 0 \tag{10b}$$

and can be chosen to be

$$(1/\alpha)[(\partial/\partial x^0)\xi^{(\alpha)}_{\mathbf{y},y_0}(\mathbf{x}, x^0)|_{x_0=y_0}] = \xi^{(\alpha)}_{\mathbf{y},y_0}(\mathbf{x}, y^0) = \delta^3(\mathbf{x} - \mathbf{y}) \tag{11}$$

where the complex number α is a degeneracy parameter. The inner products of these eigenfunctions, using the spatially local inner product eqn 6a is

$$(\xi^{(\alpha')}_{\mathbf{y}',y_0}|\xi^{(\alpha)}_{\mathbf{y},y_0}) = i\hbar(\alpha - \alpha'^*)\delta^3(\mathbf{y}' - \mathbf{y}) \tag{12}$$

and we notice that if α is real the eigenfunction has zero norm. From eqn 3, the Fourier coefficients of $\xi^{(\alpha)}_{\mathbf{y},y_0}$ are:

$$\xi^{(\alpha)}_{\mathbf{y},y_0\pm}(\mathbf{p}) = (2\pi\hbar)^{-3/2}(p^0 \pm i\alpha) \exp[(i/\hbar)(\pm p^0y^0 - \mathbf{py})] \tag{13}$$

and consequently none of these 'position' eigenfunctions lies in either the positive or negative energy subspace.

One might say that this is not *in itself* objectionable – that although the stability problem leads physics to restrict itself to positive energies, we should interpret the theory of a free system (particle or particles or fields) without regard to interactions. (As we said early in this section, Schweber (1961) seems to take this view.) And there is an obvious interpretation of why position eigenfunctions might involve energies of both signs, based on the uncertainty principle for position and momentum. Namely, the arbitrarily precise spatial confinement entails arbitrarily broad dispersion in momentum (and thereby in energy since energy is a function of momentum) and so a dispersion *greater* than the gap between $-mc$ and $+mc$. Indeed, this interpretation seems to have been proposed in the very early days of relativistic quantum theory by Landau and Peierls (1931, pp. 473–4).

So suppose we accept such position eigenstates. We will show that *still* there is a difficulty; and that two ploys to counter it run into further trouble. The difficulty arises from the need to give a probability interpretation of these eigenstates, which requires a positive definite inner product. But while X is self-adjoint and the $\xi^{(\alpha)}_{\mathbf{y},y_0}$ are mutually orthogonal in the spatially local inner product (|) of eqn 6, neither of these features hold in the positive definite inner product < | > of eqn 5, where

$$< \xi^{(\alpha')}_{\mathbf{y}'y_0} | \xi^{(\alpha)}_{\mathbf{y},y_0} >= (2\pi\hbar)^{-3} \int d^3p[(p^0/2) + (\hbar^2\alpha'^*\alpha/2p^0)] \exp[(i/\hbar)\mathbf{p}(\mathbf{y}' - \mathbf{y})]. \tag{14}$$

As a ploy to counter this difficulty, we might try defining localized states by the requirement

$$(\xi_{\mathbf{y},y_0}|\phi) := \phi(\mathbf{y}, y^0) \tag{15}$$

instead of eqn 10. With this definition we find, by comparing eqns 3 and 6c, that

$$\xi_{\mathbf{y},y_0\pm}(\mathbf{p}) = \pm(2\pi\hbar)^{-3/2} \exp[(i/\hbar)(\pm p^0y^0 - \mathbf{py})] \tag{16}$$

There is now no room for degeneracy; but, unfortunately,

$$(\xi_{\mathbf{y}',y_0}|\xi_{\mathbf{y},y_0}) = 0 \qquad \text{for all } \mathbf{y} \text{ and } \mathbf{y}' \tag{17}$$

and *all* the localized states have zero norm. Thus they are not complete.

Again, suppose we try

$$< \xi_{\mathbf{y},y_0}|\phi >:= \phi(\mathbf{y}, y^0) \tag{18}$$

Then, comparing eqns 3 and 5, we find

$$\xi_{\mathbf{y},y_0\pm}(\mathbf{p}) = (2\pi\hbar)^{-3/2} \exp[(i/\hbar)(\pm p^0y^0 - \mathbf{py})] \tag{19}$$

but now

$$< \xi_{\mathbf{y}',y_0} | \xi_{\mathbf{y},y_0} > = (2\pi\hbar)^{-3} \int (\mathrm{d}^3 p / p^0) \exp[(i/\hbar)\mathbf{p}(\mathbf{y}' - \mathbf{y})], \tag{20}$$

and the localized states are not orthogonal!

So much by way of exemplifying the difficulties of finding a position operator for the KG case. Happily, as reported at the start of this section, there *is* such an operator, with a complete set of positive energy eigenstates; viz. the NW operator. Indeed this operator is simply the projection of our naive position operator, X of eqn 9, onto the positive energy subspace. Thus one can define a Hermitian 'energy sign operator', Π, as having eigenvalues ± 1 on the positive and negative energy subspaces; the projection operators onto these subspaces are then given by,

$$\Pi_{\pm} := (1/2)(1 \pm \Pi); \tag{21}$$

and the NW operator is $\Pi_+ X \Pi_+$. (For details of this, cf. e.g. Greiner (1997). His equations 1.68, 1.163, 1.170 and 2.58 show how to write the KG equation in the form of a Schrödinger equation with Hamiltonian H, so that we can exhibit Π explicitly as $\Pi := H/\surd(H^2)$.)

Furthermore, the KG wavefunctions for an eigenstate of the NW operator have a simple Fourier integral representation: the $2p^0$ in the denominator of eqn 3 is replaced by $\surd(2p^0)$. That is, for localization at the point $\mathbf{y}$, we have (Schweber 1961, p. 62; Greiner 1997, eqn 1.179, p. 75):

$$\psi_{\mathbf{y}}(x) = \psi_{\mathbf{y}}(\mathbf{x}, x^0) = (2\pi\hbar)^{-3/2} \int (\mathrm{d}^3 p / \surd(2p^0)) \exp[(i/\hbar)(\mathbf{p}(\mathbf{x} - \mathbf{y}) - p^0 x^0)]. \tag{22}$$

But as stressed at the end of section 2, these eigenstates propagate superluminally and are delocalized under a passive Lorentz boost! Though we cannot enter into details, we can briefly state the underlying reason for the superluminal propagation: for the reason relates to the equation of motion in the positive energy subspace (regardless of the identification of position operators and localized states). This equation of evolution is not the KG equation; but instead, the first-order equation

$$i\hbar(\partial/\partial x^0)\phi(\mathbf{x}, x^0) = \surd[\kappa^2 - \hbar^2(\partial/\partial\mathbf{x})^2]\phi(\mathbf{x}, x^0). \tag{23}$$

We need to note three points. First, this *is* a formal version of the statement that we are in the positive energy subspace since the right hand side involves the operator for positive energy, $\surd[\kappa^2 - \hbar^2(\partial/\partial\mathbf{x})^2]$. Secondly, this equation supercedes the KG equation, because there is a one-to-one correspondence between positive energy solutions of the Klein–Gordon equation and specifications of (smooth) ϕ on a spacelike hyperplane and eqn 23 implies the KG equation by iteration. Third, and most important: the positive energy operator is intensely non-local, converting any function of compact support upon which it acts, into a function of unbounded support. So a positive energy wavefunction ϕ of momentarily compact support determines, via eqn 23, a time-derivative $\partial\phi/\partial x^0$ of unbounded support; and thus the wavefunction propagates instantaneously throughout all space. (This third point may seem to conflict with the fact that these wavefunctions are a subset of the set of all KG wavefunctions, which are said to propagate strictly subluminally. But it does not: for the restriction to subluminal propagation refers to wavefunctions for which *both* ϕ and its time-derivative $\partial\phi/\partial x^0$ have momentary compact support. And because of the non-locality of eqn 23, there are no such wavefunctions in the positive (or indeed, the negative) energy subspace.)

4. The Dirac wavefunction

For the Dirac wavefunction, our main point is exactly parallel to our point for the KG case: namely, once we restrict to the physically reasonable positive energy subspace, the localized states are the NW states with their (i) superluminal propagation and (ii) delocalization under a passive Lorentz-boost. Indeed, the parallels are extensive: just as for KG, naive Dirac position eigenfunctions are superpositions of positive and negative energy states, and the NW operator is simply the projection of the naive position operator onto the positive energy subspace.

There are of course some differences from the KG case. As we shall see, there is a spatially local Lorentz-invariant inner product which is positive-definite and so sustains a probability interpretation; (securing this was part of Dirac's motivation in seeking his equation). But we will end the section on yet another KG–Dirac parallel: we will review the Dirac wavefunction's difficulties about velocity and acceleration; in particular, the accelerating position operator for a free particle, the famous *Zitterbewegung*. For these difficulties also are resolved once we project to the positive energy subspace – again emphasizing our main point, that one must confront the NW states. (The KG–Dirac parallel is that velocity 'behaves well' in the KG case also, only once we project to positive energy; again, few books discuss it, but cf. Greiner (1997, pp. 70–3).)

The Dirac equation for the free evolution of a single spin-1/2 quantum particle of rest mass m, is:

$$(i\gamma^{\mu}\partial_{\mu} - \kappa)\psi(x) = 0 \tag{24}$$

where $\kappa = mc/\hbar$. The γ^{μ} are four 4×4 matrices satisfying the anti-commutation relations

$$\{\gamma^{\mu}, \gamma^{\nu}\} = 2\partial^{\mu\nu} := 2\ \text{diag}(1, -1, -1, -1) \tag{25}$$

and accordingly, the solution, $\psi(x)$, is a 4-component complex function.

Historically, Dirac was led to this 4-component formalism by searching for a locally covariant evolution equation which would support a positive definite inner product that was Lorentz-invariant and conserved in time.

The inner product

$$(\psi_2, \psi_1) := \int \mathrm{d}^3 x \psi_2^{\dagger}(x)\psi_1(x) \tag{26}$$

where $\psi^{\dagger}$ is the Hermitian adjoint of ψ (so it is a 1-row, 4-column matrix), satisfies all the desiderata. That is: for $\psi_2 = \psi_1$, the integrand is positive definite at every value of $\mathbf{x}$ and the integral is constant in time, as a consequence of eqn 24; so the integrand can be regarded as proportional to a position probability density. Furthermore the integrand is the time-component of a locally conserved 4-current,

$$j_{21}^{\mu}(x) := c\psi_2^{\dagger}(x)\gamma^0\gamma^{\mu}\psi_1(x) \tag{27}$$

the space components of which, for $\psi_2 = \psi_1$ can be regarded as the position probability current density.

The Fourier representation of $\psi(x)$, with which we can identify the probability amplitudes for momentum and energy, is given by

$$\psi(x) = (2\pi\hbar)^{-3/2} \int (\mathrm{d}^3 p/2p^0)[\tilde{\psi}_+(\mathbf{p}) \exp[(i/\hbar)(\mathbf{px} - p^0 x^0)] + \tilde{\psi}_-(p) \exp[(i/\hbar)(\mathbf{px} + p^0 x^0)] \tag{28}$$

where the four components of $\tilde{\psi}_{\pm}(\mathbf{p})$ satisfy

$$(p^0\gamma^0 - \mathbf{p} - mc)\tilde{\psi}_+(\mathbf{p}) = 0 \tag{29a}$$

and

$$(-p^0\gamma^0 - \mathbf{p} - mc)\tilde{\psi}_-(\mathbf{p}) = 0 \tag{29b}$$

As in the KG case, $\tilde{\psi}_+(\mathbf{p})$ and $\tilde{\psi}_-(\mathbf{p})$ are known as the momentum representation positive energy and negative energy state functions, respectively.

Now, how about position? At first, the naive approach seems to work well. The position probability density is given by

$$P(\mathbf{x}, x^0) := \psi^\dagger(\mathbf{x}, x^0)\psi(\mathbf{x}, x^0)/(\psi, \psi) \tag{30}$$

and the position operator by

$$[X\psi](\mathbf{x}, y^0) := \mathbf{x}\psi(\mathbf{x}, y^0) \tag{31}$$

which makes X trivially self-adjoint in the inner product. The position eigenspinors $\xi_{\mathbf{y},y_0}(\mathbf{x}, x^0)$ then satisfy

$$(\mathbf{x} - \mathbf{y})\xi_{\mathbf{y},y_0}(\mathbf{x}, x^0) = 0 \tag{32}$$

which requires

$$\xi_{\mathbf{y},y_0}(\mathbf{x}, x^0) = \delta^3(\mathbf{x} - \mathbf{y})\xi \tag{33}$$

where ξ is a constant 4-component spinor. Any 4-component spinor will do; but it can be shown that we cannot choose ξ so as to make $\xi_{\mathbf{y},y_0}(\mathbf{x}, x^0)$ an eigenspinor of the sign of the energy. (For example , this can be shown by writing the Fourier expansion 28 of the state 33 in terms of eigenspinors of helicity (defined as the component of spin in the direction of the momentum): but we skip details.) So none of these localized states 33 lies in either the positive or negative energy subspaces.

What should we make of this? First, we state two concessions. The first is familiar: as we conceded in section 3, one might hold that negative energy states in the quantum theory of a *free* system are acceptable; and one can appeal to the position–momentum uncertainty principle, to interpret why we need such states as components of localized states. Those points also apply here, in the Dirac case. The second concession is specific to the Dirac case: viz., Dirac's suggestion that we avoid the instability problem, by postulating that all but a few of the negative energy states are filled (so that the exclusion principle prevents a wholesale cascade into the negative energy 'sea') yields an argument why an arbitrarily tightly confined wave packet could not be prepared. For since the sea is filled, the negative energy states needed for such a wave packet are not available to the particle one is trying to prepare.

But our main conclusion is exactly as in the KG case. Namely: we should of course follow physics' usual restriction to the positive energy subspace; and when we do so, the position operator is the NW operator, whose eigenstates have strange properties! And as in the KG case:

(i) the NW operator is the projection of X of (4.8) onto the positive energy subspaces: it is $\Pi_+ X \Pi_+$ (where Π_+ is given explicitly by eqn 46 below);

(ii) the Dirac wavefunctions for an eigenstate of the NW operator have a simple Fourier integral representation, very similar to eqn 22. But we shall not spell out the details; (cf. Schweber (1961, pp. 94–5), Greiner (1997, pp. 279, 282)).

It will be more useful to end this section by reviewing the Dirac wave-function's difficulties about velocity and acceleration; in particular, the accelerating position operator for a free particle, the famous *Zitterbewegung*. For the fact that these difficulties disappear once we restrict to positive energy – motivating the restriction! – is not widely enough appreciated; (despite e.g. Schweber (1961, pp. 93–4) and Greiner (1997, pp. 117–20). (Nor is the fact noted at the start of this section: that in the KG case also, velocity 'behaves well' only once we project to positive energy.) One reason for this may be Dirac's own arguments (in his classic textbook) why we should accept these difficulties – arguments which we will accordingly reject!

So we turn to velocity and acceleration. The expectation value of position at time x^0 is:

$$< \mathbf{x}; x^0 >_{\psi}: = (\psi(x^0), \boldsymbol{X}\psi(x^0)) = \int \mathrm{d}^3x\psi^{\dagger}(\mathbf{x}, x^0)\mathbf{x}\psi(\mathbf{x}, x^0) \tag{34}$$

Thus

$$(d/dx^0) < \mathbf{x}; x^0 >_{\psi} = \int \mathrm{d}^3x\,\mathbf{x}(\partial/\partial x^0)(\psi^{\dagger}(\mathbf{x}, x^0)\psi(\mathbf{x}, x^0)) \tag{35}$$

and (4.1) yields, with γ the 3-vector of the γ-matrices, γ^1, γ^2 and γ^3:

$$(\partial/\partial x^0)(\psi^{+}(\mathbf{x}, x^0)\psi(\mathbf{x}, x^0)) = -(\partial/\partial\mathbf{x})(\psi^{\dagger}(\mathbf{x}, x^0)\gamma^0\gamma\psi(\mathbf{x}, \omega)) \tag{36}$$

making this substitution and integrating by parts to take $\partial/\partial\mathbf{x}$ off $(\psi^{\dagger}\gamma^0\gamma\psi)$ and put it on $\mathbf{x}$ yields:

$$(d/dx^0) < \mathbf{x}; x^0 >_{\psi} = < \gamma^0\gamma; x^0 >_{\psi} \tag{37}$$

So recalling that $x^0 = ct$, the 'velocity operator' is $c\gamma^0\gamma$: which is the speed of light times the 3-vector of Dirac matrices $\boldsymbol{\alpha}$; and since

$$(\alpha^i)^2 = (\gamma^0\gamma^i)^2 = \gamma^0\gamma^i\gamma^0\gamma^i = -\gamma^i(\gamma^0)^2\gamma^i = -(\gamma^i)^2 = 1 \tag{38}$$

it follows that the eigenvalues of any component of this 'velocity operator' are $\pm c$.

So the possible results of measuring any Cartesian component of this 'velocity' are $\pm c$! This is apparently compatible only with infinite energy and momentum; but the momentum components have as their spectrum the entire real line. Furthermore, the components of this velocity do not commute;

$$[\alpha^i, \alpha^j] = 2\delta^{ij} - 2\alpha^j\alpha^i \tag{39}$$

In successive editions of his classic textbook, Dirac lays out an argument to make sense of these light speed eigenvalues: we take the fourth edition (1958, sect. 69, p. 262). He says that velocity measurements require comparisons of positions; that for the resulting average velocity to approximate to the instantaneous velocity, we need to consider arbitrarily close times; and that such comparisons require extremely precise position measurements. These, via the position–momentum uncertainty relation, render the momentum and hence the energy so uncertain as to be dominated by the high ends of their spectra, thus pushing the resulting velocity values to that of light.

We submit that this argument is both contrived and erroneous. For first: the argument is phrased as if it applied equally to the non-relativistic case. But non-relativistically, velocity and position operators do not commute; and, despite velocity's being the derivative of position, one cannot make *precise* measurements of a given quantity by comparing *precise* measurements of another *incompatible* quantity. For the first measurement of this second quantity would destroy essential properties of the initial state pertaining to the given quantity; so that comparisons with a second measurement of this second quantity would be irrelevant to measuring the given quantity in the initial state. (Whenever something like Dirac's suggestion is actually employed, as in time of flight measurements of velocity, the individual position measurements are microscopically very imprecise, and thus allow their comparison to be informative about the velocity distribution of the initial state.)

Secondly, in the relativistic case, the velocity operator $c\boldsymbol{\alpha}$ commutes with position, and is not a function of momentum. While the first feature would seem to allow Dirac's suggested measurement procedure, the second feature undermines the appeal to the position–momentum uncertainty relation. Arbitrary uncertainty in the momentum does not influence this velocity at all!

There are further difficulties with this velocity operator. The expectation value of the velocity has a time dependence indicated by

$$< \boldsymbol{\alpha}; x^0 >_{\psi} = \int d^3x \psi^{\dagger}(\mathbf{x}, 0) e^{(i/\hbar)Ht} \boldsymbol{\alpha} e^{-(i/\hbar)Ht} \psi(\mathbf{x}, 0) \tag{40}$$

where

$$\psi(\mathbf{x}, ct) = e^{-(i/\hbar)Ht} \psi(\mathbf{x}, 0) \tag{41}$$

and the Dirac-free Hamiltonian is

$$H = c\boldsymbol{\alpha}.\mathbf{p} + \beta mc = -i\hbar c\boldsymbol{\alpha}.(\partial/\partial\mathbf{x}) + \beta mc^2. \tag{42}$$

From the anticommutation relations satisfied by the components of $\boldsymbol{\alpha}$ and β one can show (e.g. Dirac (1958, p. 263); Greiner (1997, p. 118)) that:

$$e^{(i/\hbar)Ht}\boldsymbol{\alpha}e^{-(i/\hbar)Ht} = (c\mathbf{p}/H) + (\boldsymbol{\alpha} - (c\mathbf{p}/H))e^{-2(i/\hbar)Ht} \tag{43}$$

which singles out the intuitively expected velocity operator, $\mathbf{p}c/H$, as the time-average velocity operator: that is,

$$< \boldsymbol{\alpha}; x^0 >_\psi = < \mathbf{p}c/H; 0 >_\psi + \int d^3x\psi^\dagger(\mathbf{x}, 0)(\boldsymbol{\alpha} - (\mathbf{p}c/H))\psi(\mathbf{x}, 2x^0) \tag{44}$$

and the second term oscillates with zero long time average (though the average over a time interval, T, drops to zero only as T^{-1}), the frequency spectrum satisfying $|\omega| \geq 2mc^2/h$. (Again Dirac (ibid.) defends this result as not in conflict with experimental data on velocities of free electrons. He says that actual measurements are not sensitive to these high frequency oscillations but only to the time-average constant value. To our mind, this is contrived: certainly, the insensitivity cannot be a matter of principle, for that would undermine his previous argument accounting for the eigenvalues of the velocity.)

But once we restrict to the subspaces of definite energy, velocity behaves very sensibly! To see this, we introduce, in the momentum representation, the sign-operator Π,

$$\Pi := H/\sqrt{(H^2)} = [c\boldsymbol{\alpha}.\mathbf{p} + \beta mc^2]/[c\sqrt{(\mathbf{p}^2 + m^2c^2)}] \tag{45}$$

which has eigenvalues ± 1 on the positive and negative energy subspaces, respectively. Then the energy-sign projection operators are

$$\Pi_\pm := (1/2)(1 \pm \Pi) = [E_\mathbf{p} \pm (c\boldsymbol{\alpha}.\mathbf{p} + \beta mc^2)]/2E_\mathbf{p} \tag{46}$$

where $E_\mathbf{p} = |\sqrt{(m^2c^4 + \mathbf{p}^2c^2)}|$. The projection of $c\boldsymbol{\alpha}$ into the energy-sign eigenspaces is

$$\boldsymbol{v}_\pm := \Pi_\pm c\boldsymbol{\alpha}\Pi_\pm = \pm\Pi_\pm(c\mathbf{p}/E_\mathbf{p})c = \pm(c^2\mathbf{p}/E_\mathbf{p})\Pi_\pm \tag{47}$$

The eigenstates of $\boldsymbol{v}_\pm$ are momentum eigenstates and the eigenvalues are the classically expected, $\pm(c^2\mathbf{p}/E_\mathbf{p})$, ranging in magnitude from 0 to c. So, once more, it is the subspaces of definite energy-sign that make physical sense.

But, to return to our main theme: with this restriction we have lost the naive position eigenstates – the only localized states are now the NW states, with their strange properties under time-evolution and Lorentz boosts. So we must ask the question: Is there any standpoint from which the NW localized states (and their associated position operators) make physical sense? Indeed, this question applies, not only to the Klein–Gordon and Dirac cases, but also to the arbitrary elementary systems studied by NW, and to the generalization of their concepts to non-elementary systems as well.

As announced in section 1, we believe the answer to this question is: Yes. And the rest of this essay develops such a standpoint.

5. The quantization of Lorentz covariant position variables: a prospectus

This standpoint has already been presented by one of us (GNF) in various papers, some technical (e.g. 1965a, 1965b, 1966; Fleming & Bennett 1989) and some philosophical (e.g. 1989, 1996). The core idea is that localization is always with respect to a spacelike hyperplane. There are two main novelties in our presentation in the rest of this essay. First, we emphasize classical Lorentz covariant position variables, and how these also have 'strange' properties, until we recall that localization involves a choice of spacelike hyperplane. Secondly, we emphasize the relationship between the position operators for (a) quantum field theoretic (QFT) systems, and for (b) the single particle subspaces of the QFT system's Fock space.

More precisely, our main aim will be to show how to consistently quantize classical Lorentz covariant position variables, both those for point particles and those for localizable properties of extended systems. We will proceed by trying to find self-adjoint operators that satisfy the maximum number of intuitively compelling properties suggested by classical Lorentz covariant position variables. We prefer this approach to NW's approach in terms of eigenvectors, because the position operators we obtain are not as strange in their properties as are their eigenvectors, the localized states. Of course, once the operators are obtained the properties of their eigenvectors, however strange, are fully determined and must be accepted – provided that the position operators have been constructed with sufficient generality.

First, it will be instructive to set aside a naïve notion of an 'event' locating position operator (section 6). Then we present three distinct ways of parameterizing the evolution of the Minkowski spacetime coordinates of a *classical* point particle (section 7). These parameterizations are equivalent, in the sense that one can transform back and forth between them without loss of information. One of the parameterizations, however, which we call the *hyperplane dependent parameterization* (HDP), is highly redundant and must, consequently, satisfy a constraint equation, which we call the *invariant world line condition* (IWLC). Next, in section 8, we extend our treatment from the position of a *point* particle to that of a localizable property of a spacelike extended system. We find that: (i) of the three parameterizations we considered, only the HDP remains convenient for this extension; and (ii) the IWLC must be relaxed.

Then, in section 9, we turn to quantum theory. Our procedure for quantizing the classical position variables is a 'correspondence principle': the Poincaré transformation equation for the classical position variable is taken as the transformation equation for the expectation value in a quantum state of the quantum position variable. We show that only the HDP survives quantization without serious limitations; and even it survives only in the extended form in which the quantum analogue of the IWLC is relaxed. In section 10, we give physically relevant examples; and point out the relationship between QFT position operators and single particle position operators.

We will also see (section 11) that this quantization resolves the apparent violation of Lorentz covariance encountered by NW, i.e. (ii) at the end of section 2. For we shall find that to specify the localized states, we need (apart from degeneracy due to additional degrees of freedom such as spin) three parameters in addition to those for the Minkowski coordinates of the localization – viz., parameters describing the hyperplane. Thus quantum localization takes place in a seven-dimensional manifold, rather than in four-dimensional Minkowski spacetime. From this perspective, we see that the widespread interpretation of (ii) as showing a 'frame-dependence', or even 'subjectivity', of the notion of localization, is wrong. Rather, the 3-parameter non-uniqueness of the localized states manifests a wholly objective feature of localization for *any* localizable property of a Lorentz covariant quantum system. On the other hand, we must admit that this perspective does not remove all the strange properties of the localized eigenvectors.

Finally, in section 12 we address recent criticisms of our approach, by Malament (1996) and Saunders (1994).

6. Difficulties of an 'event' locating position operator

First, it will be instructive to see the difficulties confronting the idea of an 'event' locating position operator. For it enables us to introduce two topics which will be important later: (i) our quantization procedure; and (ii) if an operator does not commute with a Casimir invariant of the Poincaré group, then it cannot be defined on an irreducible representation of the group.

The concept of a spacetime point 'event' plays a central role in classical relativity theory. One might, therefore, expect that in LIQT a 4-vector event locating position operator, X^{μ}, is useful. Different operators would be required for different types of events; e.g., the decay of a particular kind of unstable particle, and the collision of a particular set of particles, would be two types of events. The expectation value of X^{μ}, would tell us when and where, on average, an event of a given type occurs. The eigenvectors of X^{μ} would represent (unnormalizable) states in which an event definitely occurs at precise values of the Minkowski coordinates.

On the other hand, some difficulties with the notion are to be expected. First, within a given history represented by a statevector (cf. remark (6) of section 1), several events of the given type may occur; and in such a case, to which of the several events does X^{μ} refer? Secondly, since *when* the event occurs is as much a matter of observation as *where* it occurs, the time component, X^0, must be a kind of time operator; and we therefore expect at least some of the obscurity that the concept of a time operator gives rise to, even in GIQT (cf. Aharonov & Bohm 1961; 1964).

The second difficulty is more easily displayed; so we will focus on it. Let F and F' be two Minkowski coordinate systems (MCSs) related by the Poincaré transformation (Λ, a). In the spirit of the correspondence principle, we take the Poincaré transformation equation for the classical 4-vector to be the transformation equation for the expectation value of X^{μ}. So the expectation values of X^{μ}, in F and F' are related by

$$< X^{\mu} >' = \Lambda^{\mu}_{\ \nu} < X^{\nu} > + a^{\mu}. \tag{48}$$

Now let $|\Psi)$ and $|\Psi')$ represent the intrinsic history of the system in F and F' respectively, so that they are related by a unitary representation of the (Λ, a) Poincaré transformation, i.e.

$$|\Psi') = U(\Lambda, a)|\Psi). \tag{49}$$

Then we have,

$$(\Psi' X^{\mu}|\Psi') = \Lambda^{\mu}_{\ \nu}(\Psi|X^{\nu}|\Psi) + a^{\mu}(\Psi|\Psi). \tag{50}$$

Now if eqn 49 is used to replace $|\Psi')$ in the left side of eqn 50, the resulting equation holds for arbitrary $|\Psi)$; and then the linearity of X^{μ} and $U(\Lambda, a)$ justifies the removal of the statevectors, resulting in the operator transformation equation,

$$U^{\dagger}(\Lambda, a)X^{\mu}U(\Lambda, a) = \Lambda^{\mu}_{\ \nu}X^{\nu} + a^{\mu}. \tag{51}$$

For transformation parameters very close to the identity transformation, i.e. $\Lambda^{\mu}_{\ \nu} = g^{\mu}_{\ \nu} + \delta\omega^{\mu}_{\ \nu}$ and $a^{\mu} = \delta a^{\mu}$, with δ indicating a very small quantity, we have

$$U(g + \delta\omega, \delta a) = \{I + (i/\hbar)P^{\mu}\delta a_{\mu} - (i/2\hbar)M^{\mu\nu}\delta\omega_{\mu\nu} + \text{higher-order terms}\}. \tag{52}$$

where the operators P^{ν} and $M^{\nu\lambda}$ the generators of $U(\Lambda, a)$, represent the total 4-momentum and the covariant generalization of the total angular momentum of the system, respectively. Substituting eqn 52 in eqn 54 and then retaining only terms up to first order in small quantities, we obtain the commutation relations,

$$(i/\hbar)[X^{\mu}, P^{\nu}] = g^{\mu\nu}, \tag{53a}$$

and

$$-(i/\hbar)[X^{\mu}, M^{\nu\lambda}] = (g^{\mu\nu}X^{\lambda} - g^{\mu\lambda}X^{\nu}). \tag{53b}$$

But now we can state our difficulty. It follows from eqn 53a that none of the components of X commute with the operator $P^2 := P^{\mu}P_{\mu} =: M^2c^2$ that represents the square of the total rest-frame mass multiplied by the square of the speed of light. Instead, we have

$$[X^{\mu}, M^2c^2] = -2i\hbar P^{\mu}, \tag{54a}$$

or

$$[X^{\mu}, Mc] = -i\hbar(P^{\mu}/Mc). \tag{54b}$$

(To justify this division by operators, we assume there are no zero mass states, so that M has an inverse; and then use the fact that M commutes with P^{μ}.) This immediately yields the uncertainty relations,

$$\Delta X^{\mu}\Delta M c \geq (\hbar/2)| < P^{\mu}/Mc > |. \tag{55}$$

But for single stable particles, $\Delta Mc = 0$, and so the uncertainty for the event coordinates of such a system must exceed all finite bounds – with the exception of the spatial coordinates in a zero 3-momentum eigenstate. The general point here – which will recur in section 9 – is that P^2 is a Casimir invariant of the Poincaré group, and so is a multiple of the identity on any irreducible representation of the group; so that an operator not commuting with it cannot be defined on such a representation.

One might reply that this difficulty is no surprise: what (internal) event, one might ask, *could* occur in a system consisting of just one stable particle? External events are easy enough to identify; such as, when the peak of the wave packet passes a given plane. But such external events involve the prior specification of some spatial coordinates: the location of the plane.

But if this reply were right, we would expect X^μ to be unproblematic for an unstable particle. But again, there are difficulties, as follows. For such a particle, the pre-eminent event is its decay. So we might hope X^0 will pin down the time of its occurrence. But setting $\mu = 0$ in eqn 55 yields

$$\Delta X^0 \Delta Mc \geq (\hbar/2)| < P^0/Mc > | \geq (\hbar/2); \tag{56}$$

so unless ΔMc is larger than the conventional Lorentzian assessment of the spread in the rest mass spectrum of an unstable particle, the uncertainty in the time of decay will be at least as large as the decay lifetime of the particle. Furthermore, it follows from eqn 56 that, for any system whatsoever, eigenvectors of X^0 are incompatible with any finite value of rest mass dispersion. Consequently such eigenvectors must have infinite expectation values of the square of the energy.

Similarly, the application of eqn 56 to the time of collision of a two-particle scattering system seems to yield an uncertainty in the time at least as large as the time required for the individual particle wave packets to sweep over one another. Finally, from eqns 53a and 53b a non-vanishing commutator is also obtained between X^μ and the squared magnitude of the internal angular momentum (spin) of the system, yielding uncertain event coordinates when the spin magnitude is definite. We do not spell out the details here since they are somewhat more cumbersome.

To sum up: for both stable and unstable particles, an 'event' locating operator faces difficulties. (However, for an alternative approach, claiming to be grounded in quantum events, see Arshansky et al. (1983).) Accordingly, we will from now on consider localizable properties of *persistent* systems: i.e. systems many properties of which are *present at all times* (in all MCSs), and, if localizable, somewhere. For such a property it is not meaningful to enquire *when* it occurs; but only *where* it occurs at an arbitrarily given time – or, as we shall see, at an arbitrarily given appropriate generalization (in order to satisfy Lorentz covariance) of time. As announced in section 5, the next two sections consider the classical case.

7. Equivalent parameterizations of the evolution of classical position variables

In this section, we present three distinct ways of parameterizing the evolution of the Minkowski space-time coordinates of a classical point particle.

In the first parameterization, the evolution parameter is taken, in each MCS, to be the time component, x^0, of the 4-vector position variable, x^μ. The space components, x^m, $(m = 1{,}2{,}3)$ are then regarded as functions of x^0, $X^m(x^0)$. We call this *frame time parameterization* (FTP). Of our three parameterizations, this will be the one most closely related to the Galilean case.

Under a Poincaré transformation from one MCS to another,

$$x'^\mu = \Lambda^\mu{}_\nu x^\nu + a^\mu, \tag{57}$$

an FTP particle position variable transforms according to

$$X'^m(x'^0) = \Lambda^m{}_n X^n(x^0) + \Lambda^m{}_0 x^0 + a^m, \tag{58}$$

where x'^0 is, in turn, given by,

$$x'^0 = \Lambda^0{}_n X^n(x^0) + \Lambda^0{}_0 x^0 + a^0. \tag{59}$$

When (in section 9) we try to quantize this parameterization scheme, it will be useful to have the form the transformation eqn 58 takes for infinitesimal transformations, $\Lambda^\mu{}_\nu = g^\mu{}_\nu + \delta\omega^\mu{}_\nu$ and $a^\mu = \delta a^\mu$. In particular, if a first-order Taylor expansion is used to eliminate the appearance of x'^0 in the left hand side of eqn 58, we obtain (Currie, Jordan and Sudarshan 1963, eqns 3.15 to 3.17; cf. also Kerner 1972):

$$\begin{aligned} X'^m(x^0) = {} & X^m(x^0) + \delta\omega^m{}_n X^n(x^0) - \delta\omega^0{}_n(g^{mn}x^0 + X^n(x^0)[dX^m(x^0)/dx^0]) \\ & + \delta a^m - [dX^m(x^0)/dx^0]\delta a^0. \end{aligned} \tag{60}$$

This parameterization is not manifestly covariant, which one might consider to be a disadvantage. But with an eye to quantization, the frame-dependence of the parameterization seems desirable since reference to something external to the particle, viz. a coordinate system, seems to be just what is needed to allow x^0 to be *specified independently* so as to set up the circumstances under which $X^m(x^0)$ could be *measured*.

However, the relation, eqn 59, which makes the transformed evolution parameter, x'^0, dependent upon the *measured* spatial coordinates, $X^n(x^0)$, undermines the independently specified character which x'^0 also should have in the quantum theory. This potential problem appears to be ameliorated in the infinitesimal version, eqn 60, of the transformation, since a common x^0 appears throughout. But the appearance, in the right hand side of eqn 60, of the non-linear expression,

$$X^n(x^0)[dX^m(x^0)/dx^0],$$

is an expression of the same difficulty, as we shall see when quantization is attempted in section 9.

The second parameterization we consider is the simplest manifestly covariant parameterization for point particles in special relativity, and is widely discussed in textbooks. We follow convention in calling it the *proper time parameterization* (PTP). In this case a parameter, s, is introduced, invariant under Poincaré transformations, that varies along the particle's worldline in accord with the time that would be kept by a clock always at rest with respect to the particle. The particle's spacetime position is then expressed as X^{μ} (s), where,

$$[dX^{\mu}(s)/ds][dX_{\mu}(s)/ds] = 1, \tag{61}$$

$$dX^{0}(s)/ds > 0, \tag{62}$$

and under a Poincaré transformation we simply have,

$$x'^{\mu}(s) = \Lambda^{\mu}_{\ \nu}x^{\nu}(s) + a^{\mu}. \tag{63}$$

Unlike x^0, the frame-time parameter, s makes no reference to any external MCS. We can already glimpse how this will cause trouble in the quantum theory. First, it will prevent the use of s as an *independently specified* evolution parameter in the quantum theory. For how is one to *choose* the value of s for a quantum particle that was *then* to be *measured* for space-time location? Secondly, the fact that s is intrinsic threatens, through complementarity with some other intrinsic quantity such as rest energy, to make it impossible for that other quantity to have precise values. We shall see, in section 9, that this does indeed happen as a consequence of the quantum version of the transformation eqn 63. (The standard time–energy uncertainty relation does not have this kind of consequence precisely because of the extrinsic character of the frame time, x^0; (Aharonov & Bohm 1961; 1964).)

Nevertheless, at the classical level, there is no question about the equivalence of these two previous parameterizations. The formal expression of this equivalence is given by

$$X^{m}(x^{0}) = X^{m}(X^{0}(s)) = X^{m}(s), \tag{64}$$

$$x^{0} = X^{0}(s), \tag{65}$$

and

$$(1 + [dX^{m}(x^{0})dx^{0}][dX_{m}(x^{0})/dx^{0}])[dX^{0}(s)/ds]^{2} = 1. \tag{66}$$

The third parameterization we consider is constructed to contain the best features of the previous two. On the one hand, the evolution parameters (in this case there will be more than one) have, like the x^0 of FTP, an external reference to a coordinate system, and therefore, can naturally retain their independently specified character when the transition to quantum theory is made. On the other hand, the parameterization is, like PTP, manifestly covariant, thus trivially ensuring no preferred status for any one MCS.

In this parameterization (Fleming 1965a), one first *independently specifies* a metrically flat spacelike *hyperplane,* and then asks for the Minkowski coordinates of the intersection of the particle worldline with that hyperplane. The set of all spacelike hyperplanes, Σ, can be parameterized by the ordered pair, (η, τ), consisting of the future-pointing timelike unit vector, η^{μ}, normal to the hyperplane, and the timelike parameter, τ, defined so that all spacetime points lying in the hyperplane have Minkowski coordinates, $\mathbf{x}$, satisfying

$$\mathrm{x}^{\mu}\eta_{\mu} = \tau. \tag{67}$$

Under a *passive* Poincaré transformation of MCSs, F to F$'$, given by eqn 57, the parameters for any *single* hyperplane transform according to

$$\eta'^{\mu} = \Lambda^{\mu}_{\ \nu}\eta^{\nu}, \tag{68}$$

and

$$\tau' = \tau + a_{\mu}\Lambda^{\mu}_{\ \nu}\eta^{\nu}. \tag{69}$$

The particle position variable is now written as $X^{\mu}(\eta, \tau)$ and satisfies the transformation equation

$$\mathrm{X}'^{\mu}(\eta', \tau') = \Lambda^{\mu}_{\ \nu}X^{\nu}(\eta, \tau) + a^{\mu}. \tag{70}$$

We call this formulation *hyperplane-dependent parameterization* (HDP).

It is clearly redundant! This redundancy is expressed precisely by a constraint equation. First the position variable must, from eqn 67, satisfy the identity,

$$X^{\mu}(\eta, \tau)\eta_{\mu} = \tau, \tag{71}$$

which tells us that the values of $X^{\mu}(\eta, \tau)$ do, indeed, lie on the (η, τ) hyperplane.

But there is an infinity of other hyperplanes that also contain the spacetime point with the coordinates $X^{\mu}(\eta, \tau)$. And for point particles, if (η, τ) take on the values for any one of those other hyperplanes, $X^{\mu}(\eta, \tau)$ must retain the same values as for the original hyperplane. Examining this requirement for infinitesimal changes of the hyperplane parameters (η, τ) leads to the constraint equation (Fleming & Bennett 1989, eqn 4.1, p. 239),

$$[\partial X^{\mu}(\eta, \tau)/\partial \eta^{\nu}] + (X_{\nu}(\eta, \tau) - \eta_{\nu}\tau)[\partial X^{\mu}(\eta, \tau)/\partial \tau] = 0. \tag{72}$$

We call this non-linear partial differential equation relating the η and τ derivatives of X^{μ} the *invariant world line condition* (IWLC). The name stems from the fact that if eqn 72 is violated, then the worldlines traced out by X^{μ} are distinct on distinct foliations of spacetime defined by sets of parallel hyperplanes. If eqn 72 is satisfied, only one worldline is traced out, the same for every foliating subset of parallel hyperplanes.

But we will see in the next section that for HDP position variables for localizable properties of *spatially extended systems*, the IWLC can be, and usually must be, relaxed. This will be important when we turn to the quantum theory of HDP position variables.

The equivalence in the classical case, of HDP parameterization with the previous two, can now be established. For the equivalence with FTP, we simply note that for any (η, τ) there is an x^0, and for any x^0 and η there is a τ, such that

$$X^{m}(x^{0}) = X^{m}(\eta, \tau), \tag{73}$$

where, from eqn 71, the relationship between x^0 and η and τ is given by

$$x^{0} = X^{0}(\eta, \tau) = [\tau + \boldsymbol{\eta}\mathbf{X}(\eta, \tau)]/\boldsymbol{\eta}^{0} = [\tau + \boldsymbol{\eta}\mathbf{X}(x^{0})]/\eta^{0}. \tag{74}$$

In particular, for the case of $\eta = \eta^{(0)} := (1, 0, 0, 0)$, (i.e. $\boldsymbol{\eta} = 0$), we see that in each MCS the FTP position variable is conceptually identical with that particular HDP position variable for which the hyperplane is instantaneous in that MCS and is placed at the specified frame-time.

The equivalence between HDP and PTP is similar. For any future-pointing timelike unit vector, η, and any proper time, s, on the worldline, there will be a value of the inhomogeneous parameter, $\tau = \tau(\eta, \mathrm{s})$, such that

$$X^{\mu}(\eta, \tau(\eta, s)) = X^{\mu}(s). \tag{75}$$

Similarly, for any set of hyperplane parameters, (η, τ), there will be a value of the proper time on the world line, $s = s(\eta, \tau)$, such that

$$X^{\mu}(\eta, \tau) = X^{\mu}(s(\eta, \tau)). \tag{76}$$

These relations immediately establish the equivalence of the HDP and the PTP.

8. Classical violation of the invariant world line condition

We now consider the classical position variables for localizable properties of *spatially extended systems* and show that they violate IWLC.

We begin by noting that only with the HDP parameterization can one easily formulate IWLC and investigate its violation. For FTP, the basic transformation equations, 58, 59 entail IWLC; and so they would have to be relaxed, in order to investigate IWLC's violation. The reason is that in FTP the transformation equations relate a position *at an instant* in one MCS to a position *at an instant* in another MCS. From the HDP standpoint, this assumes an identity between positions on different hyperplanes belonging to distinct foliations of spacetime. But whether there is such identity – i.e. whether the traced-out worldline is independent of the foliation – is just the issue to be examined! On the other hand, for PTP it is virtually impossible to formulate IWLC at all. This is a direct consequence of the intrinsic character of the proper time parameter (i.e. its lack of reference to anything external). So, for the rest of this section we confine ourselves to the HDP.

We begin by considering a continuously distributed system characterized, for our purposes, by a local 4-vector current density, $j^\mu(x)$; which need not be conserved. We define for each hyperplane, (η, τ), a generalized charge, $Q(\eta, \tau)$, given by the integral over the hyperplane of the component of the current density normal to the hyperplane,

$$Q(\eta, \tau) := \int d^4x\delta(\eta x - \tau)j^\mu(x)\eta_\mu. \tag{77a}$$

This generalized charge is a hyperplane-dependent Lorentz scalar. That is: if a single hyperplane is represented by the parameters (η, τ) in the inertial frame F and by (η', τ') in the inertial frame F', then

$$Q'(\eta', \tau') = Q(\eta, \tau); \tag{77b}$$

expressing that the generalized charge on that hyperplane has the same value in each frame.

On any hyperplane for which this charge is non-zero, we can also define a 4-vector position variable, $X^\mu(\eta, \tau)$, which we will call the *centre of charge*. It satisfies, by construction, the defining features 70, 71 for HDP position variables. The defining equation is,

$$X^\mu(\eta, \tau)Q(\eta, \tau) := \int d^4x\delta(\eta x - \tau)x^\mu j^\lambda(x)\eta_\lambda. \tag{78}$$

It turns out, that if the current density vanishes sufficiently fast at spacelike infinity, then, by liberal use of integration by parts, the calculation of the η and τ derivatives of the right-hand sides of eqn 77a, 78 can be expressed simply. Assuming the rapid spacelike vanishing, we can show that, with $\partial j(x) := (\partial j^{\mu}(x)/\partial x^{\mu})$,

$$\begin{aligned} &Q(\eta, \tau)\{[(\partial/\partial\eta^{\nu}) + (X_{\nu}(\eta, \tau) - \eta_{\nu}\tau)(\partial/\partial\tau)]X^{\mu}(\eta, \tau)\} = \\ &-\int \mathrm{d}^4x\delta(\eta x - \tau)(x_{\nu} - X_{\nu}(\eta, \tau))j^{\mu}(x) \\ &-\int \mathrm{d}^4x\delta(\eta x - \tau)(x^{\mu} - X^{\mu}(\eta, \tau))(x_{\nu} - X_{\nu}(\eta, \tau))(\partial j(x)). \end{aligned} \tag{79}$$

But the quantity in the curly brackets on the left side of eqn 79 is just the quantity that must vanish if IWLC is to be satisfied!

We see that IWLC is not likely to be satisfied if the 4-current density has any asymmetry in its distribution about the centre of charge, or if the 4-current is not locally conserved. This would include all but very special circumstances. Furthermore, this violation of the IWLC is not mysterious. It reflects the redistribution of charge density during the evolution between mutually tilted hyperplanes. Thus non-trivial hyperplane dependence of the worldline traced out by the centre of charge for a local 4-current is a natural and well-nigh universal occurrence.

We call our next example the *centre of energy*. Again we consider a continuously distributed system, but this time characterized by its symmetric second rank stress–energy–momentum tensor density, $\theta^{\mu\nu}(x)$; which again need not be conserved, since the system need not be energetically closed. In such a case the system will have a hyperplane-dependent 4-momentum,

$$P^{\mu}(\eta, \tau) := \int \mathrm{d}^4x\delta(\eta x - \tau)\theta^{\mu\nu}(x)\eta_{\nu}, \tag{80a}$$

and covariant hyperplane-dependent angular momentum-boost tensor,

$$M^{\mu\nu}(\eta, \tau) := \int \mathrm{d}^4x\delta(\eta x - \tau)\{x^{\mu}\theta^{\nu\rho}(x)\eta_{\rho} - x^{\nu}\theta^{\mu\rho}(x)\eta_{\rho}\}. \tag{80b}$$

These quantities would be hyperplane-independent if $\theta^{\mu\nu}(x)$ were locally conserved everywhere and sufficiently rapidly vanishing at spacelike infinity.

We now need to draw a distinction between the *ordinary energy* of the system, $P^0(\eta, \tau)$, on the (η, τ) hyperplane, and the *hyperplane energy*, or, simply, energy, defined by

$$H(\eta, \tau) := P^{\mu}(\eta, \tau)\eta_{\mu}. \tag{81}$$

This energy, the timelike component of the 4-momentum normal to the hyperplane, is identical with the ordinary energy in the MCS for which the hyperplane is instantaneous; and has the virtue of being a hyperplane-dependent scalar, i.e.

$$H'(\eta', \tau') = H(\eta, \tau), \tag{82}$$

as was the charge, $Q(\eta, \tau)$ (cf. eqn 77b). Due to this scalar property we can define a centre of energy (CE) position 4-vector by

$$X^{\mu}(\eta, \tau)H(\eta, \tau) := \int d^4x\delta(\eta x - \tau)x^{\mu}\theta^{\lambda\rho}(x)\eta_{\lambda}\eta_{\rho}. \tag{83}$$

This $X^{\mu}(\eta, \tau)$ satisfies 70, 71. But if we had employed the *ordinary* energy in an analogous way to define $X^{\mu}(\eta, \tau)$ by,

$$X^{\mu}(\eta, \tau)P^{0}(\eta, \tau) := \int d^4x\delta(\eta x - \tau)x^{\mu}\theta^{0\rho}(x)\eta_{\rho} \tag{84}$$

then *that* $X^{\mu}(\eta, \tau)$ would not satisfy eqns 70, 71. It would not be a hyperplane-dependent 4-vector under Poincaré transformation. Nevertheless, for $\eta = \eta^{(0)} = (1, 0, 0, 0)$, eqns 83 and 84 yield the same values for $X^{\mu}(\eta, \tau)$. We will need this comparison presently.

Let us now examine IWLC for the CE position variable $X^{\mu}(\eta, \tau)$ of eqn 83. In the general case of an energetically open system the calculation analogous to that which led to eqn 79 in the case of the centre of charge, yields

$$\begin{aligned} H(\eta, \tau)\{[(\partial/\partial\eta^{\nu}) + (X_{\nu}(\eta, \tau) - \eta_{\nu}\tau)(\partial/\partial\tau)]X^{\mu}(\eta, \tau)\} = \varepsilon^{\mu\nu\alpha\beta}\Sigma_{\alpha}(\eta, \tau)\eta_{\beta} + \\ \int d^4x\delta(\eta x - \tau)(x^{\mu} - X^{\mu}(\eta, \tau))(x^{\nu} - X^{\nu}(\eta, \tau))(\partial\theta(x)\eta), \end{aligned} \tag{85}$$

where $\partial\theta(x)\eta := (\partial\theta^{\lambda\rho}(x)/\partial x^{\lambda})\eta_{\rho}$ and,

$$\Sigma_{\alpha}(\eta, \tau) := -(1/2)\varepsilon_{\alpha\beta\lambda\delta}(M^{\beta\lambda}(\eta, \tau) - X^{\beta}(\eta, \tau)P^{\lambda}(\eta, \tau) + X^{\lambda}(\eta, \tau)P^{\beta}(\eta, \tau))\eta^{\delta}, \tag{86}$$

is the hyperplane-dependent internal angular momentum of the system relative to its CE position variable. For closed systems, $\partial\theta(x)\eta = 0$, so that the internal angular momentum term on the right side of eqn 85 is the only source of violation of IWLC for the CE position variable.

However, for that term (unlike the case of the centre of charge), we can appeal to the familiar connection between angular momentum and rotational motion to literally *see* the violation of IWLC in a simple instance; (cf. Møller (1952, p. 170–3) and Fleming (1965a, p. 195)). Thus consider a massive spheroid, spinning counter-clockwise as seen along the axis of spin, and in the MCS in which the spheroid has no overall translational motion, i.e. its geometrical centre is at rest. Let the mass-energy of this spheroid be axially symmetrically distributed about the axis of spin. In that case the centre of *ordinary* energy will lie on the spin axis. Now relative to an MCS which is moving to the left with speed v with respect to the first MCS, the spheroid will be moving to the right (and, consequently, will not have the instantaneous shape of a spheroid). The rightward motion of the spinning ovoid will augment the instantaneous velocities of the lower half and diminish the instantaneous velocities of the upper half. It follows that, due to the velocity dependence of mass-energy, the centre of ordinary energy will lie below the geometrical centre – in the lower, more energetic half of the rotating, translating ovoid.

We stress that this is not merely a passive Lorentz transformation of the original centre of energy. The spacetime coordinates of the centre of energy in the second MCS trace out a different worldline than those of the centre of energy in the first MCS. But since the centre of hyperplane energy is coincident with the centre of ordinary energy on instantaneous hyperplanes, it follows that this *frame-dependence* of the worldline of the centre of ordinary energy is displaying the *hyperplane-dependence* of the worldline of the centre of hyperplane energy, and thus, the violation of IWLC.

Of course, the hyperplane-dependence of the worldline depends crucially on the spinning motion of the mass distribution in its overall rest-frame. We can also see this formally: the hyperplane-dependence is related to the stress–energy–momentum density, $\theta^{\mu\nu}(x)$, in a manner analogous to the relation of the first term on the right hand side of eqn 79 to the charge-current density, $j^{\mu}(x)$. This follows from a comparison of that first term on the right hand side of eqn 79 and eqn 86 and,

$$\begin{aligned} &M^{\mu\nu}(\eta,\tau) - X^{\mu}(\eta,\tau)P^{\nu}(\eta,\tau) + X^{\nu}(\eta,\tau)P^{\mu}(\eta,\tau) = \\ &\int d^4x\delta(\eta x-\tau)\{(x^{\mu} - X^{\mu}(\eta,\tau))\theta^{\nu\rho}(x)\eta_{\rho} - (x^{\nu} - X^{\nu}(\eta,\tau))\theta^{\mu\rho}(x)\eta_{\rho}\}. \end{aligned} \tag{87}$$

So we see that spatially extended rotating massive systems, upon translating, acquire asymmetries of their energy distribution. And spatial extension is guaranteed in the presence of non-zero internal angular momentum and finite mass, as Møller (ibid.) showed: he used the limiting nature of the speed of light to derive a minimum radius of such a system, namely

$$R_{\min} = S/Mc, \tag{88}$$

where S is the magnitude of the internal angular momentum and M is the mass in the overall rest-frame.

It also turns out that in the presence of non-zero S, the equal hyperplane Poisson brackets between the Minkowski components of the CE position variable do not vanish. In fact they are proportional to the left hand side of eqn 87. This is not unusual for a collective coordinate of a relativistic extended system (though as we said in remark (4) of section 1, it does restrict the role of such a coordinate in a phase space formulation).

Although there is much more that one could, with interest, examine in the HDP of classical position variables, we are now ready to turn to the quantum theoretic versions of these last two sections. The similarities will be truly remarkable! We shall find that the FTP and PTP resist quantization efforts unless one imposes restrictive and unphysical constraints; while HDP allows quantization to proceed smoothly. We shall also find that the quantum analogue of IWLC is usually violated. It is satisfied, for the position operators we consider, only in those cases where the internal angular momentum, i.e. the spin, vanishes; and even then only in the absence of most types of interactions. Finally we will find that in the cases of non-vanishing spin, the centre of 'energy' will not be identifiable with the NW position operator. It will turn out that the NW operator is best understood as a centre of the internal angular momentum distribution, a centre of spin. (And there are further similarities which we do not present. For example, Fleming (1965b, pp. 965–7) shows that the relationship between these position operators for arbitrary quantum systems, and a fourth position operator explicitly constructed to satisfy IWLC (even in the presence of spin), leads to a quantum analogue of Møller's result (eqn 88).)

9. Quantization of classical position variables

We now attempt to replace the classical position variables of section 7 by quantum position variables, i.e. by self-adjoint operators with transformation properties and evolution equations permitting their interpretation as operators representing position. Our procedure will be patterned on that used in section 6: we take the Poincaré transformation equation for the classical position variables as the transformation equation for the quantum expectation value of the position variables. So for each of our parameterizations of classical position variables, we shall try to establish the analogues of eqns 48–51.

For FTP position, the transformation equations are eqns 58, 59; so the quantum analogues of eqn 59 are ($m, n = 1, 2, 3$):

$$(\Psi'|X^m(x'^0)|\Psi') = \Lambda^m_n(\Psi|X^n(x^0)|\Psi) + (\Lambda^m_0 x^0 + a^m)(\Psi|\Psi) \tag{89a}$$

and

$$x'^0(\Psi'|\Psi') = \Lambda^0_n(\Psi|X^n(x^0)|\Psi) + (\Lambda^0_0 x^0 + a^0)(\Psi|\Psi), \tag{89b}$$

where we have emphasized the *c*-number parameter character of x^0 and x'^0 by moving them outside of the statevector inner products when possible. But it is just this c-number character that makes eqn 89 internally inconsistent. For if we (trivially) solve eqn 89b for x'^0 and substitute the expression obtained into the left side of eqn 89a and also use eqn 49 to eliminate $|\Psi')$, then the right side of eqn 89a is quadratic homogeneous in $|\Psi)$ but the left side is not unless $X^m(x'^0)$ is independent of x'^0. But these transformation equations must hold for arbitrary statevectors; and so must be consistently quadratic homogeneous on both sides or neither. Thus for FTP, we must forego a naïve correspondence principle treatment of the classical transformation equation.

Since this problem stems from the statevector dependence of the transformed frame-time parameter displayed in eqn 89b, we may be able to avoid the problem by returning to a form of the *classical* transformation equation that eliminates explicit reference to x'^0. By focusing on infinitesimal transformations, we constructed just such a form of the transformation equations in eqn 60. The expectation value version of eqn 60 is

$$\begin{aligned}(\Psi'|X^m(x^0)|\Psi') &= (\Psi|\{X^m(x^0) + \delta\omega^m_n X^n(x^0)\\ &- \delta\omega^0_n(g^{mn}x^0 + X^n(x^0) : [dX^m(x^0)/dx^0]) + \delta a^m - [dX^m(x^0)/dx^0]\delta a^0\}|\Psi),\end{aligned} \tag{90}$$

where the colon on the right hand side indicates a symmetrized product. We now substitute in the left side of eqn 90 the infinitesimal version of eqn 49

$$|\Psi') = \{I + (i/\hbar)P^\mu\delta a_\mu - (i/2\hbar)M^{\mu\nu}\delta\omega_{\mu\nu} + \text{higher-order terms}\}|\Psi), \tag{91}$$

and retain only the terms up to first order in small quantities. Removing the arbitrary statevectors and comparing terms with corresponding infinitesimal parameters, we obtain the commutation relations,

$$[X^m(x^0), P^n] = i\hbar\delta^{mn}, \tag{92a}$$

$$[X^m(x^0), P^0] = i\hbar dX^m(x^0)/dx^0, \tag{92b}$$

$$[X^m(x^0), J^n] = i\hbar\varepsilon^{mnk}X^k(x^0) \tag{92c}$$

where

$$J^n \equiv (1/2)\varepsilon^{nkl}M^{kl}, \tag{92d}$$

and

$$[X^m(x^0), M^{n0}] = i\hbar\{X^n(x^0) : dX^m(x^0)/dx^0 - x^0\delta^{mn}\}. \tag{92e}$$

It appears, then, that the quantization of frame-time parameterized classical position variables has been successfully carried out. There are two considerations, however, that should give us pause. First, it turns out that without further specification of properties of the position operator, (e.g. do the Cartesian components of the position operator commute at equal times?), the determination of the effect on the operators, $X^m(x^0)$ of a Lorentz boost is extremely complicated and not expressible in finite terms; (in particular, the naive equations, 89a, b, are no longer an option). Secondly, as was shown many years ago by Currie, Jordan, and Sudarshan (1963, pp. 364–70), if one does assume commuting components and a finite number of degrees of freedom of the physical system, then the second time derivatives, $d^2X^m(x^0)/dx^{02}$, must all equal zero. The position operators cannot accelerate! We will acquire a clearer view of this unexpected result after we discuss the quantization of HDP position variables.

We now turn to the quantization of PTP classical position variables. In this case, the intuitive transformation equation, replacing the classical position variables by the expectation values of the position operators, presents no problem. The equation in question is,

$$(\Psi'|X^\mu(s)|\Psi') = \Lambda^\mu{}_\nu(\Psi|X^\nu(s)|\Psi) + a^\mu(\Psi|\Psi). \tag{93}$$

With the exception of the insertion of the proper-time parameter, s, this equation is identical to eqn 50 and, unlike the analogous FTP eqn 89, its components are mutually consistent. Substituting eqn 49 into the left side and removing the arbitrary statevectors, $|\Psi)$, we obtain the operator equation,

$$U^\dagger(\Lambda, a)X^\mu(s)U(\Lambda, a) = \Lambda^\mu{}_\nu X^\nu(s) + a^\mu. \tag{94}$$

If we then consider infinitesimal transformations, $\Lambda = g + \delta\omega$ and $a = \delta a$, and substitute eqn 52 into the left side of eqn 94, we can extract the commutation relations,

$$[X^\mu(s), P^\nu] = -i\hbar g^{\mu\nu}, \tag{95a}$$

and

$$[X^\mu(s), M^{\nu\lambda}] = i\hbar(g^{\mu\nu}X^\lambda(s) - g^{\mu\lambda}X^\nu(s)). \tag{95b}$$

So for each value of s, our PTP position operators have the same relation to the unitary represention of the Poincaré transformations as the putative 'event' operator of section 6 did. Accordingly, the same problems arise. The position operators do not commute with the Casimir invariant P^2 of the Poincaré group, which must be a multiple of the identity on any irreducible representation of the group; so the position operators cannot be defined on such a representation. And we get e.g.

$$[X^\mu(s), Mc] = -i\hbar(P^\mu/Mc). \tag{96a}$$

which immediately yields the uncertainty relations,

$$\Delta X^\mu(s)\Delta Mc \geq (\hbar/2)| < P^\mu/Mc > |. \tag{96b}$$

Thus the discussion following eqn 55 applies, with the reference to 'event' coordinates being replaced by reference to the coordinates of the position at any given value of the proper-time s.

Returning to eqns 95a, b we notice that these commutation relations do not have the character of Heisenberg equations of motion, i.e. no derivatives of the position operators with respect to s occur on the right hand side. The mathematical reason for this is that the proper-time, s, is Poincaré invariant, $s' = s$, and so the same s appears on both sides of eqn 94. (To the observation that even though the same x^0 appeared on both sides of eqn 90 we still got Heisenberg equations of motion within the set 92a–e, the response is that the necessary derivatives were already present in eqn 90, having been present in the classical eqn 60, of which eqn 90 is the quantum analogue.)

Notwithstanding these aspects of PTP position operators, their use as dynamical variables has been extensively studied by L. Horwitz and collaborators (1983; 1988). He introduces a new basic operator, K, such that

$$[X^\mu(s), K] = -i\hbar dX^\mu(s)/ds, \tag{96}$$

for any PTP position operator. The relation of this basic operator to P^μ, and $M^{\mu\nu}$, must then be determined from additional physical hypotheses. But it remains clear from eqn 96 that PTP position operators cannot be defined on irreducible representations of the Poincaré group: so Horwitz must work with particles having unsharp rest masses.

Finally, we turn to the quantization of HDP position variables. In this case the analogue of the classical transformation eqn 70, is (Fleming 1965a, eqn 2.6)

$$(\Psi'|X^\mu(\eta', \tau')|\Psi') = \Lambda^\mu_\nu(\Psi|X^\nu(\eta, \tau)|\Psi) + a^\mu(\Psi|\Psi), \tag{97}$$

where eqns 68, 69 provide the transformation of the hyperplane parameters η and τ. The quantum analogue of the constraint eqn 71 is

$$\eta_\mu(\Psi|X^\mu(\eta,\tau)|\Psi) = \tau(\Psi|\Psi), \tag{98}$$

which, removing the arbitrary statevectors, becomes the operator constraint equation,

$$\eta_\mu X^\mu(\eta,\tau) = \tau. \tag{99}$$

Similarly, if eqn 49 is used to replace the primed statevectors on the left side of eqn 97, we can then remove the arbitrary statevectors to get the operator transformation equation,

$$U^\dagger(\Lambda,a)X^\mu(\eta',\tau')U(\Lambda,a) = \Lambda^\mu_{\ \nu}(X^\nu(\eta,\tau) + a^\mu. \tag{100}$$

For infinitesimal transformations (where $U(\Lambda, a)$ is given by eqn 52) eqn 100 yields the commutation relations,

$$[X^\mu(\eta,\tau), P^\nu] = i\hbar\{\eta^\nu(\partial X^\mu(\eta,\tau)/\partial\tau) - g^{\mu\nu}\}, \tag{101a}$$

and

$$\begin{aligned}[X^\mu(\eta,\tau), M^{\nu\lambda}] = i\hbar\{&g^{\mu\nu}X^\lambda(\eta,\tau) - g^{\mu\lambda}X^\nu(\eta,\tau)\\ &+ \eta^\nu(\partial X^u(\eta,\tau)/\partial\eta_\lambda) - \eta^\lambda(\partial X^u(\eta,\tau)/\partial\eta_\nu)\}.\end{aligned} \tag{101b}$$

Because η and τ derivatives of the position operators appear on the right hand sides of eqn 101, *these equations constitute the generalization to arbitrary hyperplanes of the Heisenberg equations of motion for position operators* (Fleming 1966, section 5).

In particular, it is important to recognize that eqn 101b is as important in this regard as is eqn 101a: eqn 101b describes the evolution of position from one hyperplane to a nearby hyperplane that is not parallel to the original, while eqn 101a describes evolution from one hyperplane to a 'nearby' parallel one. If IWLC is violated (as we have anticipated – and we shall see examples in the next section), then these modes of evolution are largely independent.

In fact, the need to handle non-trivial dynamics requires that they be independent! To see this, consider the special case of an instantaneous hyperplane, ($\eta = \eta^{(0)}$, and put, $X^m(\eta^{(0)},\tau) = X^m(x^0 = \tau)$, identifying an HDP operator with an FTP operator. If we *assume* IWLC in the operator-valued form (Fleming & Bennett 1989, eqns 4.1, 5.15),

$$[\partial X^\mu(\eta,\tau)/\partial\eta^\nu] + (X_\nu(\eta,\tau) - \eta_\nu\tau) : [\partial X^\mu(\eta,\tau)/\partial\tau] = 0, \tag{102}$$

then, using eqn 102 to replace the η derivatives in eqn 101b with $\tau = x^0$ derivatives (for the case $\eta = \eta^{(0)}$) we find that the modified eqn 101b yields the FTP commutation relation eqn 92e. But (as we mentioned in discussing eqn 92), this commutation relation, along with requiring commuting components, was found by Currie, Jordan and Sudarshan (1963, pp. 364–70) to forbid acceleration of the position operators. So enforcing IWLC, i.e. eqn 102, seems to be incompatible with non-trivial dynamical evolution. Furthermore, the discussion in section 8, where the classical IWLC *was* violated, suggests that the physical systems described by our position operators cannot be *point* particles, but, instead, must be spatially extended systems! (As mentioned at the end of section 8, cf. Fleming (1965b, pp. 965–7).)

In the case of PTP position operators, we found that systems described by these operators could not have definite rest mass. What about HDP position operators? From eqn 101a we find,

$$[X^{\mu}(\eta,\tau), P^2] = 2i\hbar\{(\eta P) : (\partial X^{\mu}(\eta,\tau)/\partial\tau) - P^{\mu}\}. \tag{103}$$

This will vanish if

$$\partial X^{\mu}(\eta,\tau)/\partial\tau = P^{\mu}/\eta P, \tag{104}$$

where the right hand side is the natural τ-velocity operator when the position operator refers to a localizable global property of the total closed system. So in that case a sharp rest mass presents no problem.

Admittedly, if the position operator refers to a property of a subsystem of the total closed system, then eqn 104 will not hold and eqn 103 will not vanish. But this simply expresses the result that for composite closed systems the observables of the subsystems are not functions of the spacetime symmetry group generators for the composite system; and so need not commute with the Casimir invariants of that symmetry group. But the subsystem in question may still have sharp rest mass. Thus if we write

$$P^{\mu} = P^{\mu}_{\mathrm{S}}(\eta,\tau) + P^{\mu}_{\mathrm{E}}(\eta,\tau), \tag{105}$$

(where the subscripts S and E stand for 'subsystem' and 'environment', including the interactions, respectively), then we need the subsystem position operator to commute with $P_{\mathrm{S}}(\eta,\tau)^2$. This would be the case if we have,

$$[X^{\mu}_{\mathrm{S}}(\eta,\tau), P^{\nu}_{\mathrm{S}}(\eta,\tau)] = i\hbar\{\eta^{\nu}(P^{\mu}_{\mathrm{S}}(\eta,\tau)/\eta P_{\mathrm{S}}(\eta,\tau)) - g^{\mu\nu}\}. \tag{106}$$

But this is just the natural condition that, in the absence of interactions, $P^{\mu}_{\mathrm{S}}(\eta,\tau)/\eta P_{\mathrm{S}}(\eta,\tau)$ be the τ-velocity operator for $X^{\mu}_{\mathrm{S}}(\eta,\tau)$. So definite rest mass of interacting subsystems is not a problem in the HDP scheme. Similar results hold for the more complex issue of definite spin.

10. Examples of HDP position operators

We begin this section by following the lead provided by our discussion of classical HDP position variables in section 8. Suppose we replace the classical, symmetric, stress–energy–momentum density field for the system of interest, $\theta^{\nu\rho}(x)$, by the quantum field theoretic (QFT) operator analogue, $\theta^{\nu\rho}(x)$. Then it follows, either from Noether's theorem (if $\theta^{\nu\rho}(x)$ is constructed from a Lagrangian density and its derivatives), or from the equal hyperplane commutation relations for $\theta^{\nu\rho}(x)$ with itself (a generalization of the equal time commutation relations for $\theta^{\nu\rho}(x)$ provided by standard QFT), that:

(1) The hyperplane-dependent tensor operators,

$$P^{\mu}(\eta, \tau) := \int d^4x\delta(\eta x - \tau)\theta^{\mu\rho}(x)\eta_{\rho}, \tag{107a}$$

and

$$M^{\mu\nu}(\eta, \tau) := \int d^4x\delta(\eta x - \tau)\{x^{\mu}\theta^{\nu\rho}(x)\eta_{\rho} - x^{\nu}\theta^{\mu\rho}(x)\eta_{\rho}\}, \tag{107b}$$

comprise, on any single hyperplane, the generators of a unitary representation of the Poincaré group. (For the special case of instantaneous hyperplanes and closed systems see, e.g., Weinberg (1995), pp. 314–18).

(2) The operators 107 also transform as hyperplane-dependent tensor operators under the unitary representation of the Poincaré group, $U(\Lambda, a)$, for the closed system consisting of both the system of interest and its environment.

(3) As a consequence of eqn 107, the operators, $X^{\mu}(\eta, \tau)$, defined (uniquely if $\eta P(\eta, \tau) := H(\eta, \tau)$ has a positive definite spectrum) by

$$X^{\mu}(\eta, \tau) : H(\eta, \tau) := \int d^4x\delta(\eta x - \tau)x^{\mu}\theta^{\nu\rho}(x)\eta_{\nu}\eta_{\rho}, \tag{108}$$

transform in accordance with eqn 100 and thus comprise HDP position operators. (To solve eqn 108 algebraically for $X^{\mu}(\eta, \tau)$ cf. Fleming (1966, eqns 7.21–23, p. 1973).)

The physical interpretation of these HDP position operators is immediate. Both from the properties of the classical analogue position variables and from the explicit form of the defining equation eqn 108, it follows that $X^{\mu}(\eta, \tau)$ locates the centre of the hyperplane energy distribution on the (η, τ) hyperplane. We will call it the centre of energy (CE) position operator.

Furthermore, the concept of such an operator is not limited to systems grounded in a field theoretic structure. For it follows from eqns 107 and 108 that $X^{\mu}(\eta, \tau)$ satisfies

$$X^{\nu}(\eta, \tau) : H(\eta, \tau) = (M^{\nu\lambda}(\eta, \tau)\eta_{\lambda} + P^{\nu}(\eta, \tau)\tau). \tag{109}$$

Eqn 109 shows that X is expressible in terms of the system's Poincaré generators. (Of course, all operators on an irreducible representation of the spacetime symmetry group are functions of the group's generators; but in a non-irreducible representation, this need not be the case.) If we now consider a general composite system consisting of many subsystems, labelled by n, the Poincaré generators for the total system and the subsystems are related by,

$$P^{\mu} = \Sigma_{n}P^{\mu}(\eta, \tau) + V^{\mu}(\eta, \tau), M^{\mu\nu} = \Sigma_{n}M_{n}^{\mu\nu}(\eta, \tau) + W^{\mu\nu}(\eta, \tau), \tag{110}$$

where $V^{\mu}(\eta, \tau)$ and $W^{\mu\nu}(\eta, \tau)$ determine the interactions between the subsystems. Consequently, if we introduce subsystem position operators via eqn 109, i.e.

$$X_{n}^{\mu}(\eta, \tau) : H_{n}(\eta, \tau) := (M_{n}^{\mu\nu}(\eta, \tau)\eta_{\nu} + P_{n}^{\mu}(\eta, \tau)\tau), \tag{111}$$

and the total closed system position operator analogously as

$$X^{\nu}(\eta, \tau) : H(\eta) := (M^{\nu\lambda}\eta_{\lambda} + P^{\nu}\tau), \tag{112}$$

(where for the closed system, $M^{\nu\lambda}$ and P^{ν} have no hyperplane-dependence, and $H(\eta)$ has no τ-dependence), we find, trivially, from eqn 110 that

$$X^{\mu}(\eta, \tau) : H(\eta) = \Sigma_{n}X_{n}^{\mu}(\eta, \tau) : H_{n}(\eta, \tau) + W^{\mu\nu}(\eta, \tau)\eta_{\nu} + V^{\mu}(\eta, \tau)\tau. \tag{113}$$

But eqn 113 sustains the centre of energy interpretation, independently of a field theoretic underpinning.

For our CE position operator, one can calculate a formal expression for the quantum IWLC, eqn 102; (the explicit evaluation requires detailed knowledge of the interaction operators $W^{\mu\nu}(\eta, \tau)$ and $V^{\mu}(\eta, \tau)$). And from eqn 109, for an arbitrary system, one can directly calculate the (equal hyperplane) commutator of the Minkowski components of the CE operator. These calculations show that the left hand side of eqn 102 is not zero, i.e. the IWLC is violated; and nor are the commutators. We discuss these in turn.

The violation of the IWLC is, perhaps, not surprising; since our discussion of the analogous classical case in section 8 showed violation for rotating spacelike extended systems. But it turns out that for closed systems, the left hand side of eqn 102 is proportional to the internal angular momentum (spin) of the system; (as are the commutators of the Minkowski components). For arbitrary systems, the violation of the IWLC includes a contribution from the spin, but also has other contributions. Does this suggest that relativistic quantum systems with non-zero spin are spacelike extended, as our classical examples were: in particular, that individual spinning 'elementary' particles, such as electrons and quarks, are spacelike extended, despite the absence of scattering data indicating internal structure for these systems? Here we merely raise the question. We hope to return to this question elsewhere.

As to the issue of non-commuting components: although they preclude there being a basis of sharply localized eigenstates on any given hyperplane (at most one component of the CE could be diagonalized), they do not undermine the physical interpretation of the CE position operator in any way: in particular, their expectation values still locate the corresponding localizable property. (However, like the non-zero Poisson brackets for the classical analogue, they complicate the relationship with phase space formulations; cf. remark (4) in section 1, and remarks following eqn 88.)

Nevertheless, to build a basis of localized states we of course need a position operator with commuting components. And as described in section 2, for elementary systems (i.e. supporting an irreducible representation space of the Poincaré group), NW found the unique such position operator, defined at a definite time i.e. on an instantaneous hyperplane. The generalization of their result to arbitrary hyperplanes is very simple and natural.

But a further and more significant generalization to *arbitrary systems,* as well as arbitrary hyperplanes, is possible – provided one augments NW's conditions (roughly, commutation of components, and eigenstates to go orthogonal under the Euclidean group; cf. section 2) with one additional condition. Namely, one demands that the generalized NW position operator be a function of the system's Poincaré generators; (just as all operators on an irreducible representation must be, and as the CE position operator was, cf. eqn 109).

Given these conditions, there is, for each system, only one position operator. The relationship between the CE and the generalized NW position operator can then also be expressed within the algebra of the system's Poincaré generators. The construction is as follows; (Fleming 1966, pp. 1972–3, 1975).

We define the system's hyperplane-dependent angular momentum by

$$J^{\mu}(\eta, \tau) := -(1/2)\varepsilon^{\mu\alpha\beta\gamma} M_{\alpha\beta}(\eta, \tau)\eta_{\gamma}, \tag{114}$$

Next, we define the system's internal angular momentum relative to the CE position operator (note: this is not quite the same as the spin – see below; cf. classical eqn 86) by

$$\Sigma^{\mu}(\eta, \tau) := J^{\mu}(\eta, \tau) + \varepsilon^{\mu\alpha\beta\gamma} X_{\alpha}(\eta, \tau)_{\beta}(\eta, \tau)\eta_{\gamma}. \tag{115}$$

We then find that

$$[X^{\mu}(\eta, \tau), X^{\nu}(\eta, \tau)] = -(i\hbar/H(\eta, \tau)^2)\varepsilon^{\mu\nu\alpha\beta}\Sigma_{\alpha}(\eta, \tau)\eta_{\beta}. \tag{116}$$

The unique generalized NW position operator, denoted by $R^{\mu}(\eta, \tau)$, is given by

$$R^{\mu}(\eta, \tau) := X^{\mu}(\eta, \tau) + (\kappa(\eta, \tau)(H(\eta, \tau) + \kappa(\eta, \tau)))^{-1}\varepsilon^{\mu\alpha\beta\gamma} P_{\alpha}(\eta, \tau)\Sigma_{\beta}(\eta, \tau)\eta_{\gamma}, \tag{117}$$

where

$$\kappa(\eta, \tau) := |(P_{\alpha}(\eta, \tau)P^{\alpha}(\eta, \tau))^{1/2}|. \tag{118}$$

This position operator has commuting components; and the internal angular momentum relative to it,

$$S^{\mu}(\eta, \tau) := J^{\mu}(\eta, \tau) + \varepsilon^{\mu\alpha\beta\gamma} R_{\alpha}(\eta, \tau)P_{\beta}(\eta, \tau)\eta_{\gamma}, \tag{119}$$

is the hyperplane generalization of the spin, in the sense that (i) its square, $S^{\mu}(\eta, \tau)S_{\mu}(\eta, \tau)$, commutes with all the system's Poincaré generators, while (ii) $S^{\mu}(\eta, \tau)$ itself satisfies the hyperplane generalization of the spin–spin commutation relation,

$$[S^{\mu}(\eta, \tau), S^{\nu}(\eta, \tau)] = i\hbar\varepsilon^{\mu\nu\alpha\beta} S_{\alpha}(\eta, \tau)\eta_{\beta}. \tag{120}$$

The relationship between the spin, $S^{\mu}(\eta, \tau)$, and $\Sigma^{\mu}(\eta, \tau)$ is given by

$$\Sigma^{\mu}(\eta, \tau) = S^{\mu}_{||}(\eta, \tau) + (\kappa(\eta, \tau)/H(\eta, \tau))S^{\mu} \perp (\eta, \tau), \tag{121}$$

where

$$S^{\mu}(\eta, \tau) := S^{\mu}_{||}(\eta, \tau) + S^{\mu} \perp (\eta, \tau), \tag{122}$$

and

$$S^{\mu}_{||}(\eta, \tau) := K^{\mu}(\eta, \tau)(K(\eta, \tau)S(\eta, \tau))/(K(\eta, \tau)^2), \tag{123}$$

and

$$K^{\mu}(\eta, \tau) := P^{\mu}(\eta, \tau) - \eta^{\mu} H(\eta, \tau). \tag{124}$$

Like the CE position operator before it, the NW position operator also violates IWLC: unless the system is closed *and* the spin is zero, in which case the two position operators are identical. One may ask: does *any* HDP position operator satisfy IWLC? In fact it is possible, for any system, to construct a position operator that: (i) in the classical limit, lies on the line joining the CE to the NW position; and (ii) when the system is closed, satisfies IWLC. However, this position operator has no special physical significance apart from the property for which it was constructed; so we will not comment further on it; (cf. Fleming 1965a, section 4).

Finally, we wish to point out an important relationship between these position operators when constructed for (a) QFT systems and for (b) single particle systems. Suppose the CE or NW position operator for a QFT system is applied to the single particle or single antiparticle sector of the QFT system's Fock state space. In that restricted domain these QFT position operators are *identical* to the single particle/antiparticle CE or NW position operators, respectively. Thus the single particle/antiparticle position operators are not independent constructions: their existence, and our study of them, in no way entails commitment to a particle theory as 'more fundamental' than quantized fields. This bears on some criticisms that will be discussed in section 12.

11. Strange properties of localized states

Having considered some physically interesting HDP position operators for arbitrary systems and some of their elementary properties, we now turn to these operators' eigenvectors, the basis vectors for localized states. We already know from sections 1 and 2 that NW localized states display two strange properties: (i) superluminal propagation; and (ii) delocalization under a passive Lorentz boost. Here we will emphasize that both properties do *not* depend on choosing the NW operator: they are generic for HDP position operators. And in line with our main aim (cf. sections 1 and 5), we will urge that sections 7 to 10 provide a standpoint from which we can make physical sense of these properties.

Note first that we will use the term 'localized state' more broadly than is customary. For us, a localized state will be any superposition of eigenvectors of a *single* Minkowski component of an HDP position operator, for which the eigenvalues lie in a domain bounded both above and below. (So a localized state will refer to a single position operator: a statevector that is localized with respect to one position operator may not be so with respect to another.) The motivation for focusing on eigenvectors of a single Minkowski component of a 4-vector operator is simply that (with the exception of the NW operator) the components of our position operators do not commute and so do not have complete sets of 'simultaneous' eigenvectors. And for a given component, eigenvalue variation within an arbitrary bounded domain conforms closely to the intuitive idea of being localized 'within' that domain; (not exactly, since such a superposition does not entail that the component has some *definite* value lying within the bounded domain).

In our notation for hyperplane dependence, the two strange properties that we need to discuss – superluminal propagation, and delocalization under a passive Lorentz boost – concern the dependence of localized states on the hyperplane parameters, τ and η respectively. We will discuss them in turn.

For τ-dependence, we will focus on superpositions of eigenvectors of a position component in a spacelike direction orthogonal to η. Thus let ξ be a spacelike unit 4-vector such that $\xi\eta = 0$, and let $|\alpha, y; \eta, \tau>$ be an eigenvector of a component HDP position operator, $Y(\eta, \tau) := \xi X(\eta, \tau)$, i.e.

$$Y(\eta, \tau)|\alpha, y; \eta, \tau> = y|\alpha, y; \eta, \tau>, \tag{125}$$

where α represents the remaining data needed to define a unique eigenvector of the usually degenerate eigenvalue y. Then

$$|\alpha, \phi; \eta, \tau> = \int dy \phi(y)|\alpha, y; \eta, \tau>, \tag{126}$$

is a state localized 'within' $\Delta \times R^2$ on the (η, τ) hyperplane, where the support of ϕ is a bounded set, Δ.

Now we are to compare two such states associated with distinct parallel hyperplanes, with domains of localization that are spacelike separated. Thus for the localized state, eqn 126, consider varying τ while holding η fixed and also replacing $\phi(y)$ with $\theta(y)$ where the support of θ is $\Delta_a := \Delta + a$. For any given change in τ, $\Delta\tau$, a sufficiently large displacement, a, can be found so that the support, Δ_a, of θ yields a localization domain, $\Delta_a \times R^2$, on the $(\eta, \tau + \Delta\tau)$ hyperplane, that is entirely spacelike separated from the localization domain, $\Delta \times R^2$, on the (η, τ) hyperplane.

Because of the spacelike separation, naïve intuition suggests that these two states should be orthogonal: i.e. the probability for finding the localizable property of the system represented by Y to be localized in the second domain, assuming localization in the first domain, should be zero. In fact, however, the second localized state is almost never orthogonal to the original localized state. That is: almost always

$$< \alpha, \theta; \eta, \tau + \Delta\tau|\alpha, \phi; \eta, \tau > \neq 0. \tag{127}$$

In this sense superluminal evolution of the position representations corresponding to components of HDP position operators is ubiquitous! This is the HDP version of the superluminal evolution extensively studied by Hegerfeldt (1974, 1985) and others in the traditional FTP formalism.

But we can now see that the original intuition is unjustifiably naïve. For general open systems, instances of superluminal evolution can even occur classically. Thus the CE for an open system can move superluminally because of independent agencies depositing energy here and removing it there. The real surprise in LIQT is that superluminal evolution happens in all *closed* systems, including individual free elementary systems!

Qualitatively, this result does not depend at all on which HDP position operator is considered, nor on the nature of the system to which the position operator refers. Quantitatively, an analysis is most easily carried out for the hyperplane generalization of the NW position operator with commuting components. In particular, if $|\alpha, x; \eta, \tau>$ is a simultaneous eigenvector of the commuting components of the NW position operator, $R^{\mu}(\eta, \tau)$, for a free elementary system, (cf. eqn 117), i.e.

$$R^{\mu}(\eta, \tau)|\alpha, x; \eta, \tau>= (x^{\mu} - \eta^{\mu}(\eta x - \tau))|\alpha, x; \eta, \tau>, \tag{128}$$

then for an arbitrary statevector of the system, $|\Psi>$, we have, with $\kappa := mc/\hbar$, an equation like eqn 21:

$$i\hbar(\partial/\partial\tau) < \alpha, x; \eta, \tau|\Psi>= [\kappa^2 - \hbar^2(\partial/\partial x)^2]^{1/2} < \alpha, x; \eta, \tau|\Psi> . \tag{129}$$

Suppose that on the (η, τ_0) hyperplane the support of $< \alpha, x; \eta, \tau_0|\Psi>$ is confined to a ball. Then writing $\tau = \tau_0 + \Delta\tau$, we find (semi-quantitatively) that for fixed $\Delta\tau$ the inner product, $< \alpha x; \eta, \tau_0 + \Delta\tau|\Psi>$, diminishes in magnitude exponentially with increasing spacelike separation from the original ball on (η, τ_0). The coefficient of exponential damping is dominated by the rest energy of the system; (Fleming & Bennett 1989, pp. 257–60).

Furthermore, the integrated total probability for finding the position eigenvalue to be spacelike separated from the original ball *decreases* with increasing $\Delta\tau$. So it is as if the system coordinate achieves maximal spacelike dispersal immediately to the future of the hyperplane of original localization; and then trickles back inside the forward light-cone envelope of the original ball.

We turn to the second strange property: delocalization under a passive Lorentz boost, i.e. what we now call the η-dependence of position eigenvectors. Michael Redhead himself touches on this property (in a paper on the vacuum in quantum field theory), for the special case of NW eigenstates for the quantized KG field. But Redhead sees this property as condemning NW localization to being 'unattractive' and 'not objective across frames' (1994, pp. 80–1). We of course hope that our account can allay his fears!

Since a spacelike hyperplane is a three-dimensional, metrically flat slice through Minkowski spacetime, the set-theoretic intersection of two intersecting spacelike hyperplanes is a two-dimensional flat slice through each hyperplane. For the (η, τ) and (η', τ') hyperplanes, with $\eta' \neq \eta$, the points in the intersection satisfy the simultaneous equations, $x\eta = \tau$, and $x\eta' = \tau'$, yielding a two-dimensional set of solutions. Now, for any HDP position operator, consider an eigenvector, $|\alpha, \tau'; \eta, \tau >$, of $\eta' X(\eta, \tau)$ such that

$$\eta' X(\eta, \tau)|\alpha, \tau'; \eta, \tau >= \tau'|\alpha, \tau'; \eta, \tau > . \tag{130}$$

Such an eigenvector is a localized state on the (η, τ) hyperplane for which the localization is confined to the points of intersection with the (η', τ') hyperplane. Similarly, the eigenvector, $|\alpha, \tau; \eta', \tau' >$, of $\eta X(\eta', \tau')$ satisfying

$$\eta X(\eta', \tau')|\alpha, \tau; \eta', \tau' >= \tau|\alpha, \tau; \eta', \tau' >, \tag{131}$$

is a localized state on the (η', τ') hyperplane with the localization again confined to the points of intersection with the (η, τ) hyperplane.

Now it is quite possible to have a single statevector, $|\Psi >$, for which both

$$< \alpha, \tau'; \eta, \tau|\Psi >= 0 \quad \text{and} \quad < \alpha, \tau; \eta', \tau'|\Psi > \neq 0 \tag{132a, b}$$

hold. In other words: on the (η, τ) hyperplane the position *could not* be found in the set of intersection points, while on the (η', τ') hyperplane it *could* be found within the same set of intersection points! Not only is this combination of results possible: if the equality eqn 132a is assumed, then the inequality eqn 132b very rarely fails! Despite the identical domain of localization for the two eigenvectors, they are very different vectors – and this allows for the strange result: that whether one can find a quantum coordinate to have values corresponding to a given set of Minkowski points can depend on which of several hyperplanes containing those points one chooses to search for the coordinate.

Just as with the superluminal propagation displayed by the τ-dependence of localized states, there is here also a classical analogue. As we saw in section 8, violation of IWLC is commonplace among classical HDP position variables for extended systems. And such violation amounts precisely to the worldline of a position variable for the foliation defined by one of two intersecting hyperplanes passing through the intersection, while the worldline for the foliation defined by the other hyperplane does not! (But we must admit that just as with superluminal τ-evolution, so here the classical analogue does not exhaustively account for the strange behaviour in the quantum case. An HDP position operator explicitly constructed to satisfy the operator version of the IWLC, eqn 102, will still display the η-evolution feature we have discussed.)

Finally, we should report, albeit briefly, our view of the main philosophical worry raised by these strange properties, especially the first. This is the worry that they could lead to causal anomalies (i.e. contradictions). Thus it seems that a particle could superluminally propagate from spacetime region A to a spacelike separated region B, where registration of the particle is used to superluminally propagate another particle into the causal past of A; where *its* registration is used to prevent, by ordinary subluminal causation, the 'release' of the first particle at A – a contradiction.

Our view is that causal anomalies must of course be avoided. But we believe they can be avoided, while accepting the above properties. We have no *proof* of this; but Fleming (1989, pp. 121–4; cf. also Fleming & Bennett 1989, pp. 259–61; Fleming & Butterfield 1992) presents arguments for it (which appeal to the second strange property, the h-dependence). Furthermore, in a sense one has no choice whether to accept these properties. For as we have shown, they are there in the formalism of LIQT: as with all problems or embarassments, ignoring them will not make them go away! This topic will come up again in section 12.

12. Reply to two recent criticisms

Finally, we would like to reply to two recent papers, by David Malament (1996) and Simon Saunders (1994) respectively, which criticize the standpoint developed in sections 5 to 11. In fact, one of us (GNF) has already replied (1996) to previous work by Saunders (1992), and by Maudlin (1994, pp. 208–12). What follows is in accord with that earlier reply. But it is worth addressing the two recent papers: because the main part of our reply is the same for each of them, and also because the reply involves expounding another important (and admittedly strange!) property – the spacelike non-commutation of HDP position projectors. We begin with that exposition.

The spectral projection for the components of any of the HDP position operators we have discussed, is given by first choosing, as for eqn 125, some ξ, a spacelike unit 4-vector such that $\xi\eta = 0$, and defining

$$Y(\eta, \tau) := \xi X(\eta, \tau). \tag{133}$$

This self-adjoint operator has a spectral decomposition,

$$Y(\eta, \tau) = \int y d\Pi_{\xi}(y; \eta, \tau), \tag{134}$$

where

$$\Pi_{\xi}(\Delta; \eta, \tau) := \int_{y \in \Delta} d\Pi_{\xi}(y; \eta, \tau), \tag{135}$$

is the projection operator onto the subspace spanned by the eigenvectors of $Y(\eta, \tau)$ with eigenvalues lying in Δ. Now consider two choices, $(\xi, \Delta, \eta, \tau)$ and $(\xi', \Delta', \eta', \tau')$, such that any spacetime point on the (η, τ) hyperplane, with Minkowski coordinates having the ξ component lying in Δ, is spacelike separated from any point on the (η', τ') hyperplane with Minkowski coordinates having the ξ' component lying in Δ'. It then turns out that, in general and counterintuitively,

$$[\Pi_{\xi}(\Delta; \eta, \tau), \Pi_{\xi'}(\Delta'; \eta', \tau')] \neq 0. \tag{136}$$

The commutator is guaranteed to vanish only for the very special case of $(\xi, \eta, \tau) = (\xi', \eta', \tau')$, i.e. same component on one and the same hyperplane.

This non-commutation gives well-known reasons for concern, both conceptual and technical. The conceptual concerns arise from the fact that a pair of non-commuting operators cannot be jointly measured precisely. This leads to two specific concerns, when the two operators are associated with spacelike separated regions.

(1) One naturally assumes that one can interpret the *association* of an operator with a spacetime region as implying that one can *measure it by performing operations confined* to that region. And once given that assumption, one wonders what could possibly prevent the joint precise measurement of two operators associated with two spacelike separated regions.

(2) If two operators do not commute, then the act of measuring one of them can influence the probabilities of outcomes of a measurement of the other. That is, measuring one operator, in accordance with Lüder's rule (the 'projection postulate') but not selecting a specific outcome, yields a state whose probability distribution for the other operator differs, in general, from that of the original state. For operators associated with two spacelike separated regions, this suggests that the act of measuring in one region can influence experimental outcomes obtained in the other region; and *that* suggests, as at the end of section 11, superluminal causation and a threat of causal anomalies. (Because the act of measuring does not select a specific outcome for the first operator, we are here confronting 'act–outcome correlations', not just 'outcome–outcome correlations': in Shimony's (1986,

pp. 146–7) terminology, now well-established in the quantum non-locality literature, we confront 'parameter dependence', not just 'outcome dependence'.)

Finally, the technical concern arises from the fact that in standard QFT, the fundamental local field operators commute (or for fermionic fields, anticommute) at spacelike separation. This makes one wonder how our spectral projectors (of HDP position operator components) are related to these field operators!

Our own response to these concerns is, briefly, as follows. As to (1): as will emerge more clearly below, we question the interpretative assumption (about interpreting 'association with a region'). As to (2): our view is as suggested by the brief discussion at the end of section 11, and by our denying the interpretative assumption. That is: we agree that causal anomalies must be avoided. But we believe – admittedly, without having a *proof*! – that they can be, while accepting not only the τ- and η-dependence of eigenstates discussed in section 11, but also the spacelike non-commutation of spectral projectors. And as to the technical concern: it turns out that once we express any of our spectral projectors as functionals of the fundamental local fields (e.g. by inserting the $X^{\mu}(\eta, \tau)$ of eqn 108 into eqn 133 and thence into eqn 135), these functionals have *unbounded* support – thus reconciling the spacelike commutation of the fundamental local fields with the non-commutation of our spectral projectors. We will see below, in discussing Saunders, that this result supports our denial of the interpretative assumption in (1).

With these points in hand, we now turn to Malament and Saunders, respectively. In a nutshell, Malament maintains that there cannot be an acceptable theory of relativistic position operators for individual quantum particles. He bases this conclusion on a theorem he proves: that any such theory must violate what we will call 'generalized microcausality', i.e. the commutativity of arbitrary physical quantities associated with spacelike separated regions. More precisely, he proves that, for $\eta = \eta'$ (i.e. parallel hyperplanes), eqn 136 is bound to occur for *any* position operator that transforms in the standard way under the translation subgroup of the Poincaré group and for which the total hyperplane energy operator, $H(\eta)$, is bounded from below. Malament then holds that his version of eqn 136 (i.e. eqn 136 restricted to parallel hyperplanes) vitiates the concept of localization, since it would threaten causal anomalies as in (2) above: so that a coherent theory of relativistic position operators for particles is impossible.

In evaluating Malament's theorem and his conclusions, it is important to realize that they do not apply only to particles. They apply equally to position operators representing localizable properties of arbitrary systems such as we have discussed. This point 'raises the ante' against rejecting such operators as physically meaningless.

But our main reply to Malament is as in our reply to (1) above. Namely: the inference to causal anomalies assumes that *association* of the projectors with spacelike separated regions involves precise joint measurability via operations *performed within* those regions. But this assumption is questionable. And once it is questioned, one can equally well interpret Malament's theorem 'contrapositively'. That is, one could argue within the standard quantum theory of measurement, that the non-vanishing commutator eqn 136 implies that precise joint spacelike separated measurements are impossible. On the other hand, *approximate* joint spacelike separated measurements that avoid causal anomalies may still be possible – and quite capable of salvaging all of our empirical data about localization. (Busch et al. (1995) is a fine recent monograph covering approximate joint measurements.) To sum up: the matter is not closed, and Malament's conclusion based on his valuable theorem is premature.

We turn to Saunders. He first reviews Malament's argument (to which he is sympathetic); he then turns to the local algebras of Algebraic Quantum Field Theory (AQFT) (Emch 1972; Haag 1992), as providing the proper framework for all discussions of localization. We will treat these themes in turn. But before doing so, we need to clarify a persistent terminological difference between him and us.

He says that in LIQT 'there is no . . . system of [localization] projections and localized states which transform covariantly, and no covariant position operator or configuration-space Born interpretation. At the level of states there is no covariant representation at all where the states have positive energy and bounded support' (1994, p. 89). He makes similar remarks elsewhere; (1994, sect. 1; cf. also Saunders 1992). And he goes on to characterize the HDP approach as relaxing covariance, in favour of a modified condition of hyperplane-dependent covariance (ibid., p. 90). We, on the other hand, have been claiming that HDP position operators and their eigenstates satisfy all legitimate covariance requirements.

The resolution of this difference lies in the fact that there is no separate condition of hyperplane-dependent covariance: rather there is *the* covariance condition applied to hyperplane-dependent quantities. Agreed, no covariant position operators and eigenvectors exist which are hyperplane-*independent*: that is the content of Saunders's claim. But hyperplane-*dependent* position operators and their eigenstates *are* covariant: this is our claim.

Both claims are correct! (Fleming (1996, pp. 21–2) expands on this point, in reply to Saunders (1992).)

Turning to substantive matters, our first point is that the early part of Saunders's paper is written as if the HDP approach to localization is based on the hope of satisfying what we called 'generalized microcausality' above; and so the paper suggests that Malament's theorem has shot the approach down. For example, Saunders says: 'it is hoped [i.e. by Fleming (1988)] that this project [i.e. hyperplane-dependent localization] could be carried through consistent with microcausality . . . Hyperplane dependence is, however, a red herring, as Malament has recently demonstrated' (p. 90).

So in reply, we want to emphasize first of all, that generalized microcausality was never an assumption, or even a 'hope', of the HDP approach. For it was clear from the beginning that generalized microcausality was violated by the HDP CE and NW operators. Admittedly, we have not presented any *systematic* treatment of how to interpret or handle the violation. But in accordance with our reply to Malament, we maintain that violation of generalized microcausality is consistent; and indeed, interpretatively coherent.

Saunders himself, embracing Malament's conclusion, turns to AQFT to seek the operator correlates of laboratory operations including those that *locate* something. Within AQFT, the association of operators with spacetime regions is determined by the support of the fundamental local fields used in the construction of the operators. Therefore, within AQFT, generalized microcausality holds as an immediate consequence of the microcausality of the fundamental fields.

But as Saunders points out, AQFT has other strange properties that thwart one's naïve intuitions about localization; (of course, *we* would say they are *as* strange as the properties of HDP localized states!). Thus Saunders reports the Reeh–Schlieder theorem (1961): which says, roughly, that any state can be approximated arbitrarily well by applying to the vacuum state (or even to an arbitrary state of bounded energy) some or other element of the algebra associated with *any* spacetime region, no matter how small. Given the assumption in (1) above (about how to interpret 'associations' of operators with regions), this means that one is able to approximate, arbitrarily closely, any state, by performing local operations, confined to any bounded region, on the vacuum state (or even on an arbitrary state of bounded energy). That is certainly hard to square with naïve, or even educated, intuitions about localization! (The same comment applies to Saunders's valuable own corollary of the Reeh–Schlieder theorem on p. 92.)

Saunders goes on to suggest that one accommodate these strange properties of localized operations, by giving up the concept of localizing *things* in favour of localizing *events*: the basic idea is that finding an event in one region does not preclude finding the same kind of event at spacelike separation. We agree that this suggestion may help one understand localization within AQFT; but we cannot pursue it here, except to remark that our section 6 seems to be relevant.

But more importantly, by way of reply to Saunders: the interpretative assumption that his discussion of AQFT seems happy to make – that locally executable precise measurements are to be represented by operators found in the local algebras – defuses the very problem that drove his discussion towards AQFT in the first place. For as we said in reply to what we called the 'technical concern' raised by eqn 136: none of the spectral projectors of the HDP position operator components is a functional of the fundamental fields with compact support. So even though a projector may be

(a) spectrally associated with a bounded region of spacetime, and
(b) interpreted in terms of localization within that region;

one *cannot* infer that the projector can be precisely measured via operations even roughly confined to the region. (And consequently, generalized microcausality cannot be required of the projectors, merely on the grounds that they are spectrally associated with spacelike separated regions – as it was by Malament.)

Finally, let us sum up our replies to Malament and Saunders, in positive terms. The violation of generalized microcausality by the HDP spectral projectors precludes the *local* execution of *precise* joint measurements of the projectors. But it may well not generate causal anomalies (which, agreed, *must* be avoided!) and it may well allow the *local* execution of *approximate* joint measurements.

Glossary of acronyms

The following acronyms are used, especially in those sections where the phrase is prevalent. We also list, where appropriate, the principal equations defining the phrase.

LIQT Lorentz-invariant Quantum Theory
GIQT Galilean-invariant Quantum Theory

NW	Newton–Wigner (eqn 117)
KG	Klein–Gordon (eqn 1)
QFT	Quantum Field Theory
MCS	Minkowski Coordinate System
FTP	Frame Time Parameterization (eqns 58, 60)
PTP	Proper Time Parameterization (eqns 61, 62)
HDP	Hyperplane Dependent Parameterization (eqns 67–70)
IWLC	Invariant World Line Condition; classical (eqn 72); quantum (eqn 102)
CE	Centre of Energy; classical (eqn 83); quantum (eqns 108, 109, 113)
AQFT	Algebraic Quantum Field Theory

References

Aharonov, Y. & D. Bohm (1961). 'Time in the Quantum Theory and the Uncertainty Relation for Time and Energy', *Physical Review* **122**, 1649–58.

(1964). 'Answer to Fock concerning the Time-Energy Uncertainty Relation', *Physical Review* **134B** , 1417–18.

Ali, S. (1998). 'Systems of Covariance in Relativistic Quantum Mechanics', *International Journal of Theoretical Physics* **37**, 365–73.

Arshansky, R., L. P. Horwitz, & Y. Lavie, (1983). 'Particles vs. Events: The Concatenated Structure of World Lines in Relativistic Quantum Mechanics', *Foundations of Physics* **13**, 1167–94.

Birrell, N. & P. Davies (1984). *Quantum Fields in Curved Space* (Cambridge: Cambridge University Press).

Busch, P., M. Grabowski & P. Lahti (1995). *Operational Quantum Physics* (Berlin: Springer Verlag).

Currie, D. G., T. F. Jordan & E. C. G. Sudarshan (1963). 'Relativistic Invariance and Hamiltonian Theories of Interacting Particles', *Reviews of Modern Physics* **35**, 350–75. [Erratum **35**, 1032.]

Emch, G. (1972). *Algebraic Methods in Statistical Mechanics and Quantum Field Theory* (New York: Wiley Interscience).

Fleming, G. N. (1965a). 'Covariant Position Operators, Spin and Locality', *Physical Review* **137B**, 188–97.

(1965b). 'Nonlocal Properties of Stable Particles', *Physical Review* **139B**, 963–68.

(1966). 'A Manifestly Covariant Description of Arbitrary Dynamical Variables in Relativistic Quantum Mechanics', *Journal of Mathematical Physics* **7**, 1959–81.

(1989). 'Lorentz Invariant State Reduction, and Localization', in A. Fine and J. Leplin (eds.), *PSA 88*, Vol. 2 (East Lansing: Philosophy of Science Association, 1989), pp. 112–26.

(1996). 'Just How Radical is Hyperplane Dependence?', in R. Clifton (ed.) *Perspectives on Quantum Reality* (Dordrecht: Kluwer, 1996), pp. 11–28.

Fleming, G. N. & H. Bennett (1989). 'Hyperplane Dependence in Relativistic Quantum Mechanics', *Foundations of Physics* **13**, 231–67.

Fleming, G. N. & J. Butterfield (1992). 'Is there Superluminal Causation in Quantum Theory', in A. van der Merwe et al. (eds.), *Bell's Theorem and the Foundations of Modern Physics* (Singapore: World Scientific, 1992), pp. 203–7.

Fulling, S. A. (1989). *Aspects of Quantum Field Theory in Curved Space-Time* (Cambridge: Cambridge University Press).

Greiner, W. (1997). *Relativistic Quantum Mechanics* (2nd edn, Berlin: Springer).

Haag, R. (1992). *Local Quantum Physics* (Berlin: Springer Verlag).

Hegerfeldt, G. (1974). 'Remark on Causality and Particle Localization', *Physical Review* **D10**, 3320–1.

(1985). 'Violation of Causality in Relativistic Quantum Theory?', *Physical Review Letters* **54**, pp. 2395–8.

Horwitz, L. P. (1983). 'On Relativistic Quantum Theory', in A. van der Merwe (ed.), *Old and New Questions in Physics, Cosmology, Philosophy and Theoretical Biology* (New York: Plenum Press, 1983), pp. 169–88.

Horwitz, L. P., R. I. Arshansky & A. Elitzur (1988). 'On the Two Aspects of Time: The Distinction and its Implications', *Foundations of Physics* **18**, 1159–93.

Itzykson, C. & J.-B. Zuber (1985). *Quantum Field Theory* (Singapore: McGraw-Hill).

Kalnay, A. J. & P. L. Torres (1973). 'Lorentz-Invariant Localization for Elementary Systems. V. General Transformations' *Physical Review* **7**, 1707–12.

Kerner, E. H. (1972). *The Theory of Action-at-a-Distance in Relativistic Particle Dynamics* (New York: Gordon and Breach).

Landau, L. & R. Peierls (1931). 'Erweiterung des Unbestimmtheitsprinzips für die relativische Quantentheorie', *Zeitschrift Physik* **69**, 56; reprinted in J. Wheeler & W. Zurek (eds.) *Quantum Theory and Measurement* (Princeton: Princeton University Press, 1983), pp. 465–76; page reference to reprint.

Landsman, N. (1999). 'Essay Review of "Quantum Mechanics on Phase Space" by F. Schroeck', *Studies in History and Philosophy of Modern Physics* **30B**, 287–305.

Lorente, M. & P. Roman (1974). 'General Expressions for the Position and Spin Operators of Relativistic Systems', *Journal of Mathematical Physics* **15**, 70–4.

Malament, D. (1996). 'In Defense of Dogma: Why There Cannot be a Relativistic Quantum Mechanics of (Localizable) Particles' in R. Clifton (ed.), *Perspectives on Quantum Reality* (Dordrecht: Kluwer, 1996) pp. 1–10.

Maudlin, T. (1994). *Quantum Non-Locality and Relativity* (Oxford: Blackwell).

Messiah, A. (1961). *Quantum Mechanics*, Vol. II (Amsterdam: North-Holland).

Møller, C. (1952). *The Theory of Relativity* (Oxford: Clarendon Press).

Newton, T. & E. Wigner (1949) 'Localized States for Elementary Systems', *Reviews of Modern Physics* **21**, 400–6.

Pauli, W. & V. Weisskopf (1934). 'Uber die quantisierung der skalaren relativistischen wellengleichung', *Helvetica Physica Acta* **7**, 709–31: reprintcd in A. Miller (ed.), *Early Quantum Electrodynamics* (Cambridge: Cambridge University Press, 1994).

Redhead, M. (1994). 'The Vacuum in Relativistic Quantum Field Theory', in D. Hull, M. Forbes & R. M. Brown (eds.), *PSA 94*, Vol. 2 (East Lansing, MI: Philosophy of Science Association, 1994), pp. 77–87.

Reeh, H. & S. Schlieder (1961). 'Bemerkungen zur unitaraquivalenz von Lorentzinvarianten Feldern', *Nuovo Cimento* **22**, 1051–68.

Saunders, S. (1992). 'Locality, Complex Numbers, and Relativistic Quantum Theory', in D. Hull, M. Forbes & K. Okruhlik (eds.), *PSA 92*, Vol. 1 (East Lansing, MI: Philosophy of Science Association, 1992), pp. 365–80.

(1994). 'A Dissolution of the Problem of Locality', in D. Hull, M. Forbes & R. M. Brown (eds.), *PSA 94*, Vol. 2 (East Lansing, MI: Philosophy of Science Association), pp. 88–98.

Schiff, L. (1968). *Quantum Mechanics* (3rd edn, Tokyo: McGraw-Hill Kogakusha).

Schweber, S. (1961). *An Introduction to Relativistic Quantum Field Theory* (New York: Harper International).

Shimony, A. (1986). 'Events and Processes in the Quantum World', in R. Penrose & C. Isham (eds.), *Quantum Concepts in Space and Time*, (Oxford: Oxford University Press, 1986); reprinted in his *Search for a Naturalistic World View*, volume II (Cambridge: Cambridge University Press); page reference to reprint.

Weinberg, S. (1995). *The Quantum Theory of Fields*, volume 1, (Cambridge: Cambridge University Press).

Wightman, A. S. (1962). 'On the Localizability of Quantum Mechanical Systems', *Reviews of Modern Physics* **34** , 845–72.

Wightman, A. S. & S. Schweber (1955). 'Configuration Space Methods in Relativistic Quantum Field Theory, I', *Physical Review* **98**, pp. 812–37.

Wigner, E. P. (1939). 'On Unitary Representations of the Inhomogeneous Lorentz Group', *Annals of Mathematics* **40**, 149-204.

(1973). 'Relativistic Equations in Quantum Mechanics', in J. Mehra (ed.), *The Physicist's Conception of Nature* (Dordrecht: Reidel), pp. 320–30.

(1983). 'Interpretation of Quantum Mechanics', in J. A. Wheeler & W. H. Zurek (eds.), *Quantum Theory and Measurement* (Princeton: Princeton University Press, 1983) pp. 260–314.

7

From metaphysics to physics

GORDON BELOT AND JOHN EARMAN

1. Introduction

Michael Redhead began his Tarner Lectures by allowing that 'many physicists would dismiss the sort of question that philosophers of physics tackle as irrelevant to what they see themselves as doing' (1995, p. 1). He argued that, on the contrary, philosophy has much to offer physics: presenting examples and arguments from many parts of physics and philosophy, he led his audience towards his ultimate conclusion that physics and metaphysics enjoy a symbiotic relationship.

By way of tribute to Michael we would like to undertake a related project: convincing philosophers of physics themselves that the philosophy of space and time has something to offer contemporary physics. We are going to discuss the relationship between the interpretative problems of quantum gravity, and those of general relativity. We will argue that classical and quantum theories of gravity resuscitate venerable philosophical questions about the nature of space, time, and change; and that the resolution of some of the difficulties facing physicists working on quantum theories of gravity appears to require philosophical as well as scientific creativity. These problems have received little attention from philosophers. Indeed, scant attention has been paid to recent attempts to quantize gravity. As a result, most philosophers have been unaware of the problem of time in quantum gravity, and its relationship to the knot of philosophical and technical problems surrounding the general covariance of general relativity – so that it has been all too easy to dismiss this latter set of problems as philosophical contrivances. Consequently, philosophical discussion of space and time has suffered.

This point is best illustrated by attending to the contrast between what philosophers and physicists have to say about the significance of Einstein's hole argument. A version of this argument was used by Earman & Norton

(1987) to argue that it is a consequence of general covariance that substantivalism about the spacetime of general relativity can be maintained only at the price of indeterminism. Philosophical responses to this version of the argument divide quite strikingly into two camps. On the one hand, there are those who criticize the argument on the grounds that it relies upon a naive approach to modality. These authors argue that the *prima facie* force of the argument evaporates once one understands the subtlety of the modal semantics of spacetime points.[1] We believe, but will not argue here, that this variety of response lacks a coherent and plausible motivation.

The second sort of response to the hole argument is more radical, and its popularity more telling as a measure of the insularity of contemporary philosophy of space and time. The modalists acknowledge that the hole argument has *some* value: it points up a strange fact about the modal semantics of spacetime theories. A more radical response is to deny that the hole argument has *anything* at all to teach us about the nature of spacetime.[2] This is often combined with a general pessimism concerning the present state of philosophical discussion of space and time. Thus, Rynasiewicz contrasts the present state of the debate with its glorious past:

> What is remarkable about the substantival-relational debate is that, although it engaged natural philosophers from the seventeenth century into the nineteenth century and continues to be debated in academic philosophy, interest in the controversy on the part of twentieth century physicists has waned over the generations to virtually nil. (1992, p. 588)

Meanwhile, Leeds questions the interest of interpretative work on general relativity by contrasting the hole argument literature with another genre of philosophy of physics:

> There is an oddity here, it seems to me: for surely the philosophers of physics who work on these problems are the same men and women who, in another mood, are fond of comparing quantum mechanics with GTR, as the paradigm case of a theory which cries out for interpretation with the paradigm case of a theory which does not. (1995, p. 428)

1 This camp subdivides into two factions: those who attempt to derive an appropriately sophisticated modal semantics for spacetime from some general framework (Bartels 1996; Brighouse 1994; Butterfield 1989; and Maudlin 1990); and those who take the required semantics as a primitive (Hoefer 1996; Maidens 1993; and Stachel 1993).

2 Again there are two factions. On the one hand, we have those who hold the argument is fallacious because determinism is a formal property of theories, independent of questions of interpretation (Leeds 1995 & Mundy 1992). On the other hand, we have those who claim that the hole argument turns upon a piece of philosophy of language (the inscrutability of reference), and has nothing to do with philosophy of physics (Liu 1996 & Rynasiewicz 1996).

Discussion of the hole argument is often taken to be the epitome of irrelevant philosophy of physics. It is held, implicitly or explicitly, that it is *obvious* that there is nothing to the argument, since no physicist would entertain for a minute the proposition that general relativity is an indeterministic theory. It is supposed to be something of an embarrassment that philosophers have wasted so much time on this argument – how do they expect physicists to take them seriously? Typically, partisans of this line of thought believe that the hole argument is predicated on some sort of simple mistake. Most spectacularly, it is claimed that it has nothing in particular to do with general relativity at all. Rather, it is an artifact of a certain misguided way of thinking about language, naively mistaken for a bit of philosophy of physics:

> Such permutation arguments have been exploited at length by W. V. Quine, Donald Davidson, and Hilary Putnam to argue that a hankering for absolute criteria of individuation leads to an inscrutability of reference. The hole argument is nothing more than an application of the same techniques to space-time theories. If it yields relationist or anti-realist conclusions, these are conclusions which apply globally to any ontology. The substantival-relational debate, however, was a local one over the status of space and time.[3]

It is further alleged that philosophy of space and time has been led yet further astray by the suggestion that the 'solution' to the hole argument is to be found in the furthest reaches of metaphysics:

> I think that issues about whether *this* spacetime point could in some other world have been over *there* are not really questions about the nature of spacetime points, or indeed about physics at all, they are questions about *situations* or *possible worlds* – philosophers' constructions so loosely connected with reality that we can consistently answer these questions in any way that we care to. And in fact it seems to me that this had begun to be the consensus about these questions until Earman seemed to breathe new life into them via the connection with determinism. (Leeds 1995, p. 436)

These philosophers of physics paint a bleak picture indeed of the current state of the philosophy of space and time: having long ago lost its relevance to physics, it has recently degenerated into the worst sort of confused and eminently *philosophical* discussion.

This is in sharp contrast to the interest in the substantival–relational debate which is expressed by some physicists:

[3] Rynasiewicz (1996), p. 305. See also pp. 243–44 of Liu (1996) and p. 84 of Maudlin (1989).

> I would like to argue that the problem of quantum gravity is an aspect of a much older problem, that of how to construct a physical theory which could be a theory of an entire universe and not just a portion of one. This problem has a long history. It was, I believe, the basic issue behind the criticisms of Newtonian mechanics by Leibniz, Berkeley, and Mach. (Smolin 1991, p. 230)

This remark is not atypical: many physicists who work on canonical quantum gravity believe that the substantival–relational debate is directly relevant to their research.[4] In fact, many physicists emphasize the importance of interpretative questions about general relativity – often motivated by the belief that differences of opinion about the technical and conceptual difficulties of quantum gravity can be traced to differences of opinion concerning the classical theory. Thus, Rovelli asserts that

> many discussions and disagreements on interpretational problems in the quantum domain (for instance the famous 'time issue') just reflect different but unexpressed interpretations of the *classical* theory. Thus, the subtleties raised by the attempts to quantize the theory force us to reconsider the problem of observability in the classical theory. (1991c, pp. 297–8)

Furthermore, far from dismissing the hole argument as a simple-minded mistake which is irrelevant to understanding general relativity, many physicists see it as providing crucial insight into the physical content of general relativity. Thus, Isham uses a version of the hole argument to motivate an important claim about the observables of classical and quantum gravity:

> the diffeomorphism group moves points around. Invariance under such an active group of transformations robs the individual points of $\mathcal{M}$ of any fundamental ontological significance . . . This is one aspect of the Einstein 'hole' argument that has featured in several recent expositions (Earman & Norton 1987; Stachel 1989). It is closely related to the question of what constitutes an *observable* in general relativity – a surprisingly contentious issue that has generated much debate over the years and which is of particular relevance to the problem of time in quantum gravity. In the present context, the natural objects are Diff($\mathcal{M}$)-invariant spacetime integrals . . . Thus the 'observables' of quantum gravity are intrinsically non-local. (Isham 1993, p. 170)

Most surprisingly, one can even find physicists grappling with issues about the transworld identification of spacetime points:

[4] This belief is much less common among string theorists.

> The basic principles of general relativity – as encompassed in the term 'the principle of general covariance' (and also 'principle of equivalence') – tell us that there is no natural way to identify the points of one space-time with corresponding spacetime points of another.[5]

In short, a survey of the literature on quantum gravity reveals a very different picture of the relevance of philosophical work on the nature of space and time from that which is current among philosophers of physics. We do not mean to suggest that physicists are universally enthusiastic about the substantival–relational debate in general, or about the hole argument in particular. Nor, of course, are all philosophers of physics ill-disposed towards these topics. What *is* true is that philosophers of physics have tended to be unaware of the extent of the interest which physicists take in these issues. Philosophy of physics has suffered as a result: the interpretative inter-relationship between classical and quantum gravity has been missed; and the interest of the questions surrounding the general covariance of general relativity has been underestimated.

Our purpose in this short essay, is to bring these shortcomings to the attention of philosophers of physics, and to begin to redress them by giving a brief outline of the relationship between the interpretative problems of classical and quantum gravity, as we understand it.[6] Our focus in this paper is the canonical approach to quantum gravity, in which general relativity is first cast in Hamiltonian form, and then quantized via the canonical procedure. Unfortunately, the Hamiltonian formulation of general relativity is not entirely straightforward. Rather than being a true Hamiltonian system, general relativity is a gauge theory – and a somewhat peculiar one at that. The first task, undertaken in section 2, is to describe this formalism, and how its peculiarities derive from the general covariance of the standard formulation of general relativity. We will also discuss the interpretative problems of general relativity *qua* gauge theory. In the following section, we will see that the problem of time in quantum gravity – surely one of the deepest conceptual problems facing contemporary physics – follows from the gauge invariance of general relativity. Along the way, we will attempt to explicate the relationship between the somewhat unfamiliar conceptual problems of canonical gravity, and familiar philosophical problems about the nature of space, time, and change.

[5] Penrose (1996), p. 591. Penrose is led to this conclusion by the same considerations that motivate Rovelli and Isham; see especially p. 586.

[6] We discuss these issues at greater length in a companion paper, Belot & Earman (1999).

2. General relativity as a gauge theory

A Hamiltonian system consists of a phase space equipped with a real-valued function, H, the *Hamiltonian*. The geometric structure of the phase space is such that the specification of the Hamiltonian determines a unique curve $t \rightarrow x(t)$, called a dynamical trajectory, through each point of the space (see figure 7.1). Ordinarily, one thinks of the points of phase space as representing the dynamically possible states of some classical physical system, of the Hamiltonian as encoding information about the energy of each of these states, and the dynamical trajectory through a given point of phase space as representing the unique dynamically possible past and future of the state represented by that point. Thus interpreted, a Hamiltonian system constitutes a complete and deterministic description of a classical system.

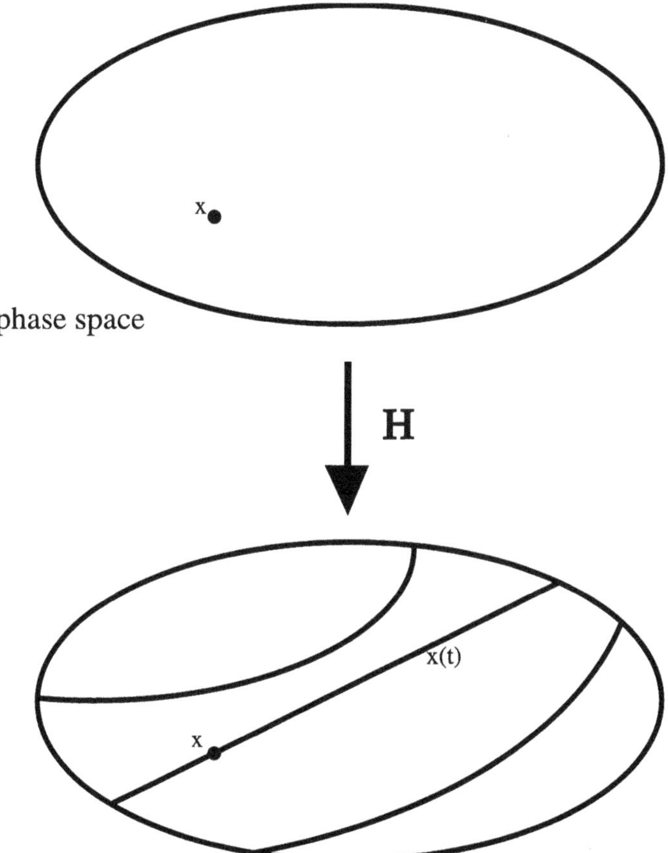

Figure 7.1 *Hamiltonian systems*

Unfortunately, the most natural formulations of many interesting classical theories – including electrodynamics and general relativity – are not strictly Hamiltonian. Rather they are gauge theories, in which the equations of motion fail to uniquely determine the evolution of the state in phase space. The geometric structure of the phase space of a gauge theory is somewhat weaker than that of a Hamiltonian system. For our purposes, the most important point is that the phase space of a gauge theory is naturally foliated by submanifolds of some fixed dimension, called gauge orbits (see figure 7.2). It is convenient to introduce the following notation and terminology: if x is a point in phase space, then $[x]$ is the unique gauge orbit in which x lies; if x and y are points of phase space then $x \sim y$ iff $[x]=[y]$; if f is a function on phase

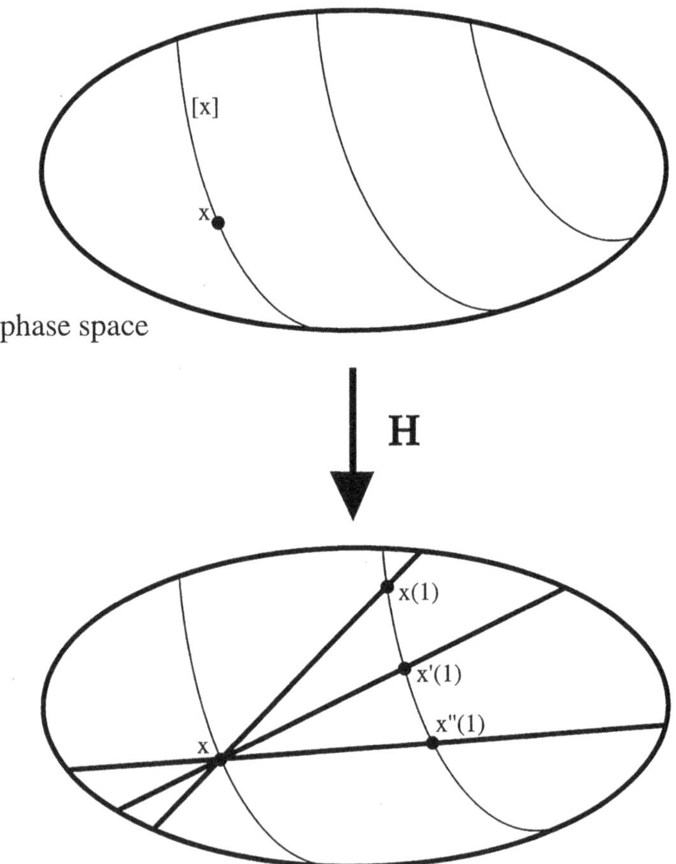

Figure 7.2 *Gauge systems.*

space which is constant on gauge orbits – i.e. if $x \sim y$ implies $f(x) = f(y)$ – then f is said to be *gauge invariant.* As in the Hamiltonian case, a gauge theory consists of a phase space equipped with a real-valued Hamiltonian H (we require that H be gauge invariant). But whereas in the former case the Hamiltonian together with the geometry of the phase space determined a unique dynamical trajectory through each point, in the gauge-theoretic case we find that there are infinitely many dynamical trajectories through each point of the phase space. Thus, if we fix an initial state, x, our theory is unable to tell us which point of phase space represents the state of the system at some later time t_1 – since we can find distinct dynamical trajectories, $x(t)$ and $x'(t)$, through our initial point x_0, and in general we expect that $x(t_1 \neq x'(t_1)$. What makes gauge theories interesting, however, is that although they are incapable of predicting which point of phase space represents the future state of the system, they *do* predict which gauge orbit that point will lie in – we find that $x(t_1) \sim x'(t_1)$ even if $x(t_1) \neq x'(t_1)$ (see figure 7.2).

The gauge freedom inherent in the equations of motion of a gauge theory complicates interpretation. It is possible to adopt the same *literal* approach which works so well for Hamiltonian systems, according to which each point of phase space corresponds to exactly one dynamically possible state. In this case, however, there will be physically real quantities which are not gauge invariant, since points lying in the same gauge orbit will correspond to distinct physically possible states of the system.[7] This straightforward approach has a serious disadvantage: it renders the theory indeterministic. Indeed, the state represented by our initial point, x_0, will have many possible futures: if $x(t)$ and $x'(t)$ are distinct dynamical trajectories through this point, then $x(t_1)$ and $x'(t_1)$ represent distinct physically possible future states of the system at time t_1. Note, however, that this indeterminism need not render the theory empirically inadequate: one can maintain that all *observable* quantities are gauge invariant, so that the theory can still be used to make determinate predictions of measurement outcomes, even if it cannot determine the evolution of all physical quantities.

Alternatively, we can require that our interpretation be *gauge invariant*, in the sense that all physically real quantities are represented by gauge invariant functions on phase space. In this case, points of phase space which lie in the same gauge orbit will correspond to the same dynamical state, and determinism will be rescued (since $x(t_1) \sim x'(t_1)$). *Ceteris paribus*, this sort

[7] Here and below we assume that two states are distinct if there is some physically real quantity which takes on different values in the two states.

of interpretation will be preferable to a literal interpretation, since we prefer to think of our classical theories as being deterministic.

All of this can be illustrated using the most familiar gauge theory, electrodynamics. Here the phase space is the set, $\{(A, E)\colon A, E\colon S \to \mathbb{R}^3, \text{div } E = 0\}$, of vector potentials and electric fields on physical space, S. The gauge orbits have the following structure: $(A, E) \sim (A', E')$, iff $E' = E$ and $A' = A + \text{grad } \Lambda$, for some $\Lambda\colon S \to \mathbb{R}$. The Hamiltonian for vacuum electrodynamics is $H = \int_S |E|^2 + |\text{curl } A|^2 dx^3$, and the equations of motion are $\dot{A} = -E$ and $\dot{E} = \text{curl}(\text{curl } A)$. These equations determine the evolution of E uniquely, but determine the evolution of A only up to the addition of the gradient of a scalar. That is, if $(A(t), E(t))$ and $(A'(t), E'(t))$ are two dynamical trajectories through the same initial point in phase space, then for all t we have that $E'(t) = E(t)$ and $A'(t) = A(t) + \text{grad } \Lambda(t)$, for some scalar Λ. If we give this theory a literal interpretation by stipulating that A corresponds to the velocity field of a material ether, then the theory becomes indeterministic – according to $A(t_1)$ *this* bit of ether ends up *here*, while according to $A'(t_1)$ it ends up *there*.[8] On the other hand, we can stipulate that along with the electric field, E, the only other physically real quantity is the magnetic field, $B \equiv \text{curl } A$. The theory is rendered deterministic, since B is a gauge invariant quantity whose evolution is determined uniquely by the equations of motion.[9] The vector potential becomes a mathematical fiction, and any information contained in A over and above that contained in B is, to borrow Redhead's apt phrase, *surplus structure* (see Redhead 1975).

We now turn to the more complicated case of general relativity. The points of the phase space of general relativity should represent instantaneous states of the gravitational field. Thus, it is natural to build the phase space out of points which represent the geometries of Cauchy surfaces of models of general relativity. We proceed as follows. We fix a three-dimensional manifold, Σ. We now imagine that this manifold is embedded in a model, (M, g), of general relativity as a Cauchy surface. What geometrical information does Σ inherit from (M, g)? The answer is: the first and second fundamental forms, h and k, of Σ considered as a submanifold of (M, g). Here h is just the Riemannian metric which results from restricting g to Σ, and k is, *very* roughly, the time derivative of h. Thus we can think of h and k as being the position and momentum variables of our gravitational field theory. The phase space of general relativity is just the set of

[8] Such an interpretation would, presumably, be supplemented with an account of measurement which would imply that this indeterminism would be empirically undetectable.

[9] See Belot (1998) for an account of the difficulties which this interpretation faces in light of the Aharonov–Bohm effect.

pairs, (h, k), which can arise from embedding Σ as a Cauchy surface of a model of general relativity. The gauge orbits of this phase space have a strikingly simple structure: (h, k) and (h', k') lie in the same gauge orbit iff they can be viewed as Cauchy surfaces of the *same* model of general relativity.

Thus, if we fix a model of general relativity and look at the geometries, (h, k) and (h, k'), corresponding to two distinct Cauchy surfaces, then these geometries will lie in the same gauge orbit of general relativity. This means that the dynamical trajectory which joins these two points must lie entirely within their common gauge orbit. It follows that in general relativity the Hamiltonian assumes a very simple form: $H \equiv 0$. This is, in fact, a straightforward consequence of the general covariance of the theory: in order to have a non-zero Hamiltonian, one must have access to a preferred parameterization of time (see chapter 4 of Henneaux & Teitelboim 1992 for a careful discussion). This feature, that the dynamical trajectories are restricted to gauge orbits, rather than passing from one orbit to another, distinguishes general relativity from other familiar gauge theories and is the source of some of the most interesting interpretative problems of classical and quantum gravity.

Indeed, once general relativity has been formulated as a gauge theory, we can reformulate the hole argument so that it depends upon the structure of the phase space of general relativity rather than upon the diffeomorphism invariance of the standard formulation of the theory. Above, we noted if $x(t)$ and $x'(t)$ are two dynamical trajectories passing through the same initial point, x_0, then these trajectories represent distinct dynamical futures for x_0 under a literal interpretation, but represent the same dynamical future under a gauge invariant interpretation. Thus, general relativity, like any gauge theory, is indeterministic under a literal interpretation, and deterministic under a gauge invariant interpretation. But notice that in general relativity $x(t)$ and $x'(t)$ correspond to the same four-dimensional geometry of spacetime (since they each represent a sequence of instantaneous geometries which belong to the same gauge orbit). This is the core of the hole argument, re-expressed in the language of gauge theories.

The connection with the substantival–relational debate can be recovered as follows. First, notice that in general relativity $x(t)$ and $x'(t)$ correspond to the same four-dimensional geometry. Next, consider the condition which Leibniz and Clarke agreed constituted a good criterion for distinguishing absolutists from relationalists: the absolutist will affirm, while the relationalist will deny, that there could be two worlds whose contents instantiated the same spatial relations, but which were numerically distinct in virtue of the fact that different points of space would be occupied by the material objects of the two worlds. In the context of general relativity, the natural generalization of this criterion is: substantivalists will affirm, while relationalists will deny, that there could be two general relativistic worlds which instantiated the same four dimensional geometry which were numerically distinct in virtue of the fact that the geometrical relations of these worlds would be differently shared out among the spacetime points of the worlds.[10] Thus, substantivalists will view $x(t)$ and $x'(t)$ as representing distinct instantiations of the given four-geometry by a set of existent spacetime points. Relationalists, on the other hand will maintain that all instantiations of a given four-dimensional geometry are numerically identical – $x(t)$ and $x'(t)$ correspond to the same physical possibility. Hence relationalism is a gauge invariant interpretation of general relativity.

At this point it would seem to be mandatory to adopt a gauge invariant interpretation of general relativity. Otherwise, we are committed to ruling general relativity to be indeterministic for the slimmest of reasons: a metaphysical preference for substantivalism. Certainly, there are a number of prominent gravitational physicists who accept this line of thought (see, e.g., Rovelli 1991c and 1997). Most philosophers who have written on the hole argument concur (although they are more likely to opt for some sophisticated form of substantivalism than for Rovelli's robust relationalism). In the next section, we will discuss the bearing that considerations arising out of quantum gravity have on this question. Our conclusion will be that, when the dust settles, these considerations may well override any grounds for settling the substantival–relational dispute which are internal to general relativity itself. Before turning to this argument, however, we would like to point out that, even at the classical level, the formulation of a cogent gauge invariant interpretation of general relativity is by no means a straightforward task.[11]

[10] For the purposes of this paper, we bracket the question of the cogency of the sort of substantivalists, mentioned briefly in 7.1, who deny the Leibniz–Clarke condition; see Earman (1989) and Belot (1999). For present purposes, it suffices to observe that *some* varieties of substantivalism are literal interpretations of general relativity.

[11] Many philosophers of physics have assumed that the formulation of a gauge-invariant interpretation of general relativity is an easy and attractive option – simply count diffeomorphic models as physically equivalent. We will see below, however, that this strategy runs into serious technical and conceptual difficulties when we attempt to formulate interpretations in terms of the phase space of the theory rather than individual models. The upshot for the substantival–relational debate remains unclear. But the existence of these problems, and the fact that they do not arise for other gauge theories such as electromagnetism, makes it wholly implausible that the issues surrounding the hole argument are pseudo-problems or merely a recapitulation of familiar problems about reference.

Our first worry is of a technical nature. Recall that the problem of isolating the gauge invariant quantities of a theory is closely related to the problem of formulating a gauge invariant formulation of that theory. Indeed, according to a gauge invariant interpretation all physically real quantities are represented by gauge invariant quantities on phase space. It is thus impossible to fully specify a gauge invariant interpretation until the gauge invariant functions have been identified. In the case of general relativity, this is quite a tall order. Very few gauge invariant quantities are known. Worse, in the spatially compact case it has been proven that there are no gauge invariant quantities which are local (i.e., which can be written as integrals over Σ of h, k, and a finite number of their derivatives; see Torre 1993). Thus, there is reason to worry that the gauge invariant quantities of general relativity may not be suitable candidates for the ontology of an interpretation of a classical field theory like general relativity.

Thus, it is not at all trivial to formulate a gauge invariant interpretation of general relativity. As long as the existence of a sufficient number of suitable gauge invariant quantities of general relativity remains an open question, a dark cloud hangs over the programme of giving a gauge invariant interpretation of the theory. We contend that an honest approach to the interpretative enterprise requires one to suspend judgement until these difficult technical questions are settled. In support of this contention, we note that the isolation of the gauge invariant quantities of the theory appears to be a prerequisite for a gauge invariant approach to quantization. And, we claim, an interpretation which supports quantization is deeper than one which does not. We conclude that the present state of ignorance about the existence of gauge invariant quantities for general relativity should give pause to advocates of gauge invariant interpretations.[12]

Our second problem is primarily conceptual in nature, and prefigures the problems to be discussed in the next section. Although it lies near the core of the conceptual difficulties facing attempts to quantize gravity, it is simple enough to state: *prima facie, gauge invariant interpretations of general relativity imply that time and change are illusions.* Lest it be thought that this is a pseudo-problem concocted by benighted philosophers, let us begin with a formulation from a leading gravitational physicist:

> How can changes in time be described in terms of objects which are completely time independent? In particular, since the only physical, and thus

[12] This is closely related to the claim, advanced in Earman (1989), that relationalists are obliged to produce formulations of physical theories which can be expressed in relationally pure vocabulary.

> measurable quantities are those which are time independent, how can we describe the rich set of time dependent observations we make of the world around us? (Unruh 1991, p. 266)

The argument here is straightforward. Fix a model of general relativity, (M, g), with two Cauchy surfaces, Σ_1 and Σ_2. If we accept that the only physically real quantities of general relativity are gauge invariant, then it follows that there is no physically real quantity which takes on different values when evaluated on Σ_1 and Σ_2. This is to say that there is no change in the world described by (M, g), since no physically real quantity evolves in time. *Prima facie*, proponents of gauge invariant interpretations of general relativity are committed to the view that change is illusory. This is a very radical thesis – it is, for instance, much stronger than the doctrine that there is no room for temporal becoming in a relativistic world, since proponents of the tenseless theory of time are confident that they can account for the existence of change (see, e.g., Mellor 1981).

There has been a great deal of discussion of this thesis, and its consequences, in the physics literature. It is clear that at the classical level it does not affect the routine business of applying general relativity – there is no difference of opinion as to the predictions of general relativity between *Parmenideans* who hold that time and change are illusory and the *Heraclitians* who believe that change is a fundamental reality. None the less, the question of the reality and nature of time and change is the subject of debate among physicists (see, e.g., the exchange between Kuchař and Rovelli on pp. 138–40 of Ashtekar & Stachel 1991). The reason for this is straightforward: it is felt that one must correctly understand the nature of time and change in the classical theory if one is to make any progress on the deep conceptual problems of quantum gravity.

3. Quantum gravity and the problem of time

In the previous section, we saw that it was possible to formulate general relativity as a gauge theory. This cast the interpretative problems of general relativity in a new light: it became clear that the hole argument is just a special case of the general observation that literal interpretations of gauge theories are indeterministic. This observation motivates us to search for gauge invariant interpretations of general relativity. But this turns out to be unexpectedly difficult: unsolved technical problems and daunting conceptual difficulties stand in our way. The latter, especially, are troubling:

adopting a gauge invariant interpretation of general relativity seems to require us to revise out most fundamental metaphysical categories. Thus we find ourselves in the following position: having noticed that general relativity is a gauge theory, we attempt to apply the interpretative strategy which works so well for other gauge theories, such as electrodynamics. But here the peculiar nature of general relativity *qua* gauge theory – the fact that the Hamiltonian is zero, so that dynamical trajectories are restricted to gauge orbits – forces us to confront difficulties that have no analogues in other familiar gauge theories. The case is similar when we attempt to quantize general relativity. Here the most obvious strategy is to apply to general relativity the algorithm of canonical quantization which works so well for other gauge theories. In this section we will see that this leads to tremendous conceptual difficulties. These may be traced back, via the vanishing of the classical Hamiltonian, to the general covariance of general relativity.

The algorithm for quantizing gauge theories is simple enough in outline.[13] As in ordinary quantum mechanics, one begins by selecting a set of classical position and momentum variables, and then constructing a representation of their algebra as an algebra of operators on the Hilbert space $L^2(Q, \mu)$ (here Q is the classical configuration space and μ is some appropriate measure on Q). One then isolates the subspace, $\mathcal{H}$, of $L^2(Q, \mu)$ consisting of gauge invariant wave functions. Once equipped with a suitable inner product, $\mathcal{H}$ will be the space of states for our quantum theory. In order to complete the construction, we need to introduce an algebra of gauge invariant quantum observables on $\mathcal{H}$, and a quantum Hamiltonian, $\hat{H}$, which determines the dynamics of the theory via the Schrödinger equation.

In the case of electrodynamics, we take the components of A and E at each point of physical space, S, to be our classical position and momentum variables. We then construct $L^2(Q, \mu)$ by taking Q to be the set, $\{A\colon S \to \mathbb{R}^3\}$, of vector potentials. The elements of $L^2(Q, \mu)$ are wave functions over Q: complex functionals of the form $\Psi[A]$. We construct $\mathcal{H}$ by restricting our attention to those $\Psi \in L^2(Q, \mu)$ which are gauge invariant in the sense that $A \sim A'$ implies $\Psi[A] = \Psi[A']$. We then find a set of self-adjoint operators on $\mathcal{H}$ which represent an algebra of gauge invariant classical observables, and impose a quantum version of the classical Hamiltonian, $H = \int_S |E|^2 + |\mathrm{curl\ A}|^x 3$. The result is a quantum theory of

[13] The details, of course, involve many subtleties, which we gloss over in the following. It is safe to assume that the quantization of gravity faces all of the technical difficulties present in other quantum field theories – operator ordering ambiguities, anomalies, problems of regularization and renormalization, etc. – and then some.

the familiar type: a Hilbert space carrying a self-adjoint representation of an algebra of observables, with dynamics given by a Schrödinger equation.

It is possible, at least formally, to apply this algorithm to general relativity.[14] One takes the components of h and k at each point of Σ as the classical position and momentum variables. Thus, our states will be wavefunctions on $Q = \text{Riem}\ \Sigma$, the space of Riemannian metrics on Σ. We restrict our attention to gauge invariant wavefunctions.[15] The observables must remain unspecified for the time being, since so few classical gauge invariant quantities have been identified. But it is easy to write down the correct quantum Hamiltonian: $\hat{H} \equiv 0$. Thus, the Schrödinger equation becomes trivial:

$$\frac{\partial \Psi[h]}{\partial t} = i\hbar \hat{H} \Psi[h] = 0.$$

Since the quantum Hamiltonian is zero, there is no evolution in time of the quantum states. This is the core of the *problem of time*: there appears to be no time or change in quantum gravity. This is not surprising: the algorithm sketched above for the construction of quantum gravity treats the gauge invariance of general relativity in strict analogy with the gauge invariance of other theories. And we know from the discussion of section 2 that embracing an interpretation of general relativity which is gauge invariant in this sense involves, at least *prima facie*, renouncing the existence of time and change. But, whereas in the classical domain our interpretative beliefs did not interfere with our ability to apply the theory, it appears to be impossible to understand this gauge invariant theory of quantum gravity as a theory about *our* world, replete as it is with change. For, naively applying a fragment of the conceptual apparatus of ordinary quantum mechanics, it appears that the states, $\Psi[h]$, of our theory of quantum gravity tell us that the probability of measuring the spatial geometry to be (Σ, h) at a given instant is given by $|\Psi[h]|^2$. But according to our theory the state never evolves, and so the probability of obtaining a given three-geometry as the outcome of a measurement is constant in time. But this contradicts one of

[14] I.e. although technical problems – such as the prohibitive difficulty of constructing an appropriate measure on the configuration space of general relativity – prevent one from achieving a fully rigorous formulation of quantum gravity along these lines, it is possible to formally manipulate the equations. Even at this level, one can see that the resulting theory of quantum gravity would run into serious interpretative problems.

[15] Naively, we might hope that the gauge invariant wavefunctions would be those for which $\Psi[h] = \Psi[h']$ if h and h' can be viewed as spatial geometries of Cauchy surfaces of the same model. Unfortunately, the actual situation is considerably more complicated. See pp. 189–93 and 225–6 of Isham (1993).

the most basic tenets of modern cosmology: that the geometry of the universe is temporally evolving.

Thus, the most straightforward approach to the construction of a quantum theory of gravity results in a theory which appears to be incapable of describing our world. In recent years, many attempts have been made to remedy this situation: either by showing that this appearance is misleading and that one can indeed construct a viable theory of quantum gravity by treating the gauge invariance of general relativity in analogy with the gauge invariance of other theories; or by suggesting alternative routes to quantum gravity which rely upon unorthodox interpretations of the gauge invariance of general relativity.[16] We will briefly describe four such attempts, and sketch their interpretative underpinnings. Although we will not go into details, it is safe to assume that each of these proposals is fraught with serious technical and conceptual difficulties (see Belot & Earman 1999 for further details and references).

One of the most radical proposals is Barbour's timeless interpretation of quantum gravity (see Barbour 1994a, b). Barbour accepts the Parmenidean interpretation of general relativity, and the approach to quantum gravity sketched above to which it leads. He also endorses the account of measurement according to which the probability of finding the spatial geometry (Σ, h) is given by $|\Psi[h]|^2$. He must, therefore, face the full force of the problem of time. His response is to bite the bullet: Barbour acknowledges that he is committed to the view that the probability of measuring (Σ, h) does not change in time; his explanation for this surprising result is that *there is no time*. On his view, what exists is a single moment, and a wavefunction which tells us the probabilities for possible outcomes of measurements of the geometry of this instant. The geometries which are likely measurement outcomes are supposed to encode information which would make it appear *as if* the universe had a past and future. But in fact, on Barbour's view, all that exists, and all that we experience, is a single magical moment.

Rovelli has developed a somewhat less radical Parmenidean approach to quantum gravity (Rovelli 1991a, b). He begins by endorsing the Parmenidean approach to general relativity, and the relationalism which underlies it. Thus, he posits that the physically real quantities in general relativity are gauge invariant. These are constants of motion of the theory, in the sense that they take on the same value at any two Cauchy surfaces of a given model of general relativity. Thus, if we are talking about a system

[16] Kuchař (1992) is the canonical survey.

containing a rocket, 'the mass of the rocket' will not be a physically real quantity, since it changes over time. But quantities of the form 'the mass of the rocket at blast-off' and 'the mass of the rocket when it docks at the space-station' *will* be constants of motion – they take on a single value for each model of the theory. Rovelli's insight is that we can give the set of constants of motion of the theory an internal structure by grouping together constants of motion: we can form an 'evolving constant of motion' whose members are just the constants of motion which give the mass of the rocket at each instant. The resulting set will form a one-parameter family. We can even write down an equation which describes the change in the value of the evolving constant as the parameter is varied. Rovelli's hope is that we can do the same at the quantum level: group the quantum observables which correspond to classical constants of motion into quantum evolving constants. This is a technically daunting task. If it can be carried out, then one could hope to write down equations which would govern the evolution in parameter time of the expectation values of the quantum observables. One would then have explained how the appearance of time can arise out of fundamentally timeless structures. This would not, however, amount to 'finding time' in quantum gravity, since Rovelli doubts that there is a unique time hidden here. Rather, one expects that there will be many ways of constructing evolving constants, classical and quantum. We should be opportunistic about selecting a technique which suits the model at hand, and our psychological experience of time, without reading our pragmatic decision back into nature.

On the Heraclitian side, we again find a very radical approach which privileges spatial structure, and a less radical, but more ambitious proposal. The former consists of breaking the general covariance of general relativity by introducing a privileged time parameter.[17] Using this preferred time coordinate, we can rewrite general relativity as a time-dependent Hamiltonian system with a non-zero Hamiltonian. It is then in principle possible to construct a quantum field theory of gravity in which the privileged classical time parameter provides the background for the evolution of the quantum states. Thus, general relativity and quantum gravity are recast as theories of the evolution of the gravitational field in time. In particular, general relativity becomes a theory of the evolution in time of the geometry of space. One ends up with theories of gravity which are no more difficult

[17] This can be done in a number of ways. Popular approaches include the introduction of special forms of matter and the privileging of the foliations of spacetime by Cauchy surfaces of constant mean curvature. See Kuchař (1992) for references and discussion.

(or easy!) to interpret than other classical and quantum field theories. Of course, this is achieved at the price of sacrificing one of the conceptual cornerstones of contemporary physics: the idea that the spirit behind the general covariance of general relativity forbids one from introducing preferred coordinate systems.

Kuchař's internal time proposal represents a more plausible Heraclitian alternative (see Kuchař 1972 and 1993). He endeavours to respect the spirit of the general covariance of general relativity without treating it as a principle of gauge invariance. His starting point is the conviction that if one wants to make sense of our experience of change, then one must accept that there are physically real quantities of general relativity and quantum gravity which are not gauge invariant. As argued in section 2, such a position is closely associated with substantivalism about the spacetime of general relativity. Kuchař's goal is to isolate within the classical phase space some structure which deserves to be called temporal, but which would not single out a preferred time parameter.[18] This temporal structure would allow one to explicate a sense in which the physically real quantities of general relativity evolve, and hence cannot be gauge invariant. One would then use the same temporal structure as the background against which the states of quantum gravity would exhibit non-gauge invariant evolution. Kuchař's programme is very ambitious technically, and is, as yet, incomplete. But it suggests a way in which substantivalism, despite its shortcomings, can underwrite a distinctive and intriguing approach to the quantization of general relativity.

4. Conclusion

In the previous section, we saw how the problem of time in quantum gravity arises out of the conceptual difficulties surrounding the general covariance of general relativity. We sketched four proposed solutions to the problem of time, and saw how each was linked to a definite view about the nature of change and time in the classical and quantum world, and to a view about the nature of the spacetime of general relativity. One expects, of course, that each of these programmes, if developed rigorously, would lead to a different theory of quantum gravity. If one of them should turn out to be empirically adequate, that fact would have interpretative repercussions at the classical

[18] The vagueness here is, of course, our own: we gloss over the details of Kuchař's sophisticated and elegant construction.

level – if an interpretation of general relativity suggests a given approach to quantization, then one is bound to revise one's interpretative judgements should that approach prove untenable. Thus, we find the interpretative problems of general relativity and quantum gravity to be bound in a close relationship: we cannot settle one set of questions without this having repercussions for the other set. And so long as the way forward in quantum gravity is unclear, physicists will continue to ponder and to debate metaphysical questions about the nature and existence of spacetime and change. Thus, we reach the same conclusion as Michael did in his Tarner lectures: 'that physics and metaphysics blend into a seamless whole, each enriching the other, and that in very truth neither can progress without the other' (Redhead 1995, p. 87).

References

Ashtekar, A. & J. Stachel (eds) (1991). *Conceptual Problems of Quantum Gravity* (Boston: Birkhäuser).

Barbour, J. (1994a). 'The Timelessness of Quantum Gravity: I. The Evidence from the Classical Theory', *Classical and Quantum Gravity* **11**, 2853–73.

(1994b). 'The Timelessness of Quantum Gravity: II. The Appearance of Dynamics in Static Configurations', *Classical and Quantum Gravity* **11**, 2875–97.

Bartels, A. (1996). 'Modern Essentialism and the Problem of Individuation of Spacetime Points', *Erkenntnis* **45**, 25–43.

Belot, G. (1998). 'Understanding Electromagnetism', *British Journal for the Philosophy of Science* **49**, 531–55.

(1999). 'Rehabilitating Relationalism', *International Studies in Philosophy of Science* **13**, 35–52.

Belot, G. & J. Earman (1999). 'Pre-Socratic Quantum Gravity', forthcoming in C. Callender & N. Huggett (eds.) *Philosophy Meets Physics at the Planck-Scale* (Cambridge: Cambridge University Press).

Brighouse, C. (1994). 'Spacetime and Holes', in D. Hull, M. Forbes & R. M. Burian (eds.) *PSA 1994*, Vol. 1 (East Lansing, MI: Philosophy of Science Association), pp. 117–25.

Butterfield, J. (1989). 'The Hole Truth', *British Journal for the Philosophy of Science* **40**, 1–28.

Earman, J. (1989). *World Enough and Space-Time* (Cambridge, MA: MIT Press).

Earman, J. & J. Norton (1987). 'What Price Spacetime Substantivalism? The Hole Story', *British Journal for the Philosophy of Science* **38**, 515–25.

Henneaux, M. & C. Teitelboim (1992). *Quantization of Gauge Systems* (Princeton: Princeton University Press).

Hoefer, C. (1996). 'The Metaphysics of Space-Time Substantivalism', *Journal of Philosophy* **XCIII**, 5–27.

Isham, C. (1993). 'Canonical Quantum Gravity and the Problem of Time', in L. A. Ibort and M. A. Rodriguez (eds.) *Integrable Systems, Quantum Groups, and Quantum Field Theories* (Dordrecht: Kluwer, 1993), pp. 157–288.

Kuchař, K. (1972). 'A Bubble-Time Canonical Formalism for Geometrodynamics', *Journal of Mathematical Physics* **13**, 768–81.

(1992). 'Time and Interpretations of Quantum Gravity', in G. Kunsatter, D. Vincent and J. Williams (eds.) *Proceedings of the 4th Canadian Conference on General Relativity and Astrophysics* (Singapore: World Scientific, 1992), pp. 211–314.

(1993). 'Canonical Quantum Gravity', in R. J. Gleiser, C. N. Kozameh and O. M. Moreschi (eds.) *General Relativity and Gravitation 1992: Proceedings of the Thirteenth International Conference on General Relativity and Gravitation held at Cordoba, Argentina, 28 June–4 July 1992* (Philadelphia: Institute of Physics Publishing), pp. 119–50.

Leeds, S. (1995). 'Holes and Determinism: Another Look', *Philosophy of Science* **62**, 425–37.

Liu, C. (1996). 'Realism and Spacetime: Of Arguments Against Metaphysical Realism and Manifold Realism', *Philosophia Naturalis* **33**, 243–63.

Maidens, A. (1993). '*The Hole Argument: Substantivalism and Determinism in General Relativity*', Ph.D. thesis, University of Cambridge.

Maudlin, T. (1989). 'The Essence of Space-Time', in M. Forbes & A. Fine (eds) *PSA 1988*, Vol. 2 (East Lansing MI: Philosophy of Science Association), pp. 82–91.

(1990). 'Substances and Space-Time: What Aristotle Would Have Said to Einstein', *Studies in History and Philosophy of Science* **21**, 531–61.

Mellor, D.H. (1981). *Real Time* (Cambridge: Cambridge University Press).

Mundy, R. (1992). 'Space-Time and Isomorphism', in D. Hull, M. Forbes & K. Okruhlik (eds.) *PSA 1992*, Vol. 1 (East Lansing, MI: Philosophy of Science Association), pp. 515–27.

Penrose, R. (1996). 'On Gravity's Role in Quantum State Reduction', *General Relativity and Gravitation* **28**, 581–600.

Redhead, M. (1975). 'Symmetry in Intertheory Relations', *Synthese* **32**, 77–112.

(1995). *From Physics to Metaphysics* (Cambridge: Cambridge University Press).

Rovelli, C. (1991a). 'Is There Incompatibility Between the Way Time is Treated in General Relativity and in Standard Quantum Mechanics?', in A. Ashtekar & J. Stachel (eds.) *Conceptual Problems of Quantum Gravity* (Boston: Birkhäuser, 1991), pp. 126–40.

(1991b). 'Time in Quantum Gravity: An Hypothesis', *Physical Review D* 43, 442–56.

(1991c). 'What is Observable in Classical and Quantum Gravity', *Classical and Quantum Gravity* **8**, 297–316.

(1997). 'Halfway Through the Woods: Contemporary Research on Space and Time', in J. Earman & J. Norton (eds.) *The Cosmos of Science* (Pittsburgh: University of Pittsburgh Press, 1997), pp. 180–223.

Rynasiewicz, R. (1992). 'Rings, Holes and Substantivalism: On the Program of Leibniz Algebras', *Philosophy of Science* **59**, 572–89.

(1996). 'Absolute versus Relational Space-Time: An Outmoded Debate?', *Journal of Philosophy* **XCIII**, 279–306.

Smolin, L. (1991). 'Space and Time in the Quantum Universe', in A. Ashtekar & J. Stachel (eds.) *Conceptual Problems of Quantum Gravity* (Boston: Birkhäuser, 1991), pp. 228–91.

Stachel, J. (1989). 'Einstein's Search for General Covariance, 1912-1915', in D. Howard and J. Stachel (eds.) *Einstein and the History of General Relativity* (Boston: Birkhäuser, 1989), pp. 63–100.

(1993). 'The Meaning of General Covariance', in J. Earman, A. Janis, G. Massey and N. Rescher (eds.) *Philosophical Problems of the Internal and External Worlds: Essays on the Philosophy of Adolph Grünbaum* (Pittsburgh: University of Pittsburgh Press, 1993), pp. 129–60.

Torre, C. G. (1993). 'Gravitational Observables and Local Symmetries', *Physical Review D* **48**, R2373–R2376.

Unruh, W. (1991). 'No Time and Quantum Gravity', in R. Mann & P. Wesson (eds.) *Gravitation: A Banff Summer Institute* (Singapore: World Scientific, 1991), pp. 260–75.

8

Models and mathematics in physics: the role of group theory

STEVEN FRENCH

1. Introduction

The relationship between mathematics and science is clearly of fundamental concern in both the philosophy of mathematics and the philosophy of science. How this relationship should be represented is a crucial issue in this area. One possibility is to employ a model-theoretic framework in which 'physical' structures are regarded as embedded in 'mathematical' ones. In section 2 I will briefly outline a form of this type of account which offers a function space analysis of theories (Redhead 1975).[1] This function space analysis is then used to represent the relationship between theoretical and mathematical structures. In subsequent sections I will consider the role of group theory in physics from within this meta-theoretical framework and then draw some conclusions for realism in the philosophy of science.

2. Function spaces and the model-theoretic approach

According to Redhead, it is an 'empirical-historical fact' that theories in physics can be represented as mathematical structures[2] (Redhead 1975).

Much of the content of this essay arose out of discussions with Otávio Bueno, Sarah Kattau and James Ladyman as part of a series of informal workshops on group theory and physics at the University of Leeds. I am grateful to them and to Koji Tanaka both for their specific comments and their indulgence. I would also like to thank Jeremy Butterfield and the referee of this volume for further helpful suggestions. The blame for any idiocies in the work lies entirely with me, of course.

[1] As Pagonis has indicated (1996), this can be construed as a version of the model-theoretic approach.

[2] There is a slight ambiguity here as shortly after Redhead refers to the empirical-historical fact that 'physical theories are always *related* to mathematical structures' in the way he has subsequently spelt out (1975, p. 89; my emphasis).

This then allows the possibility of representing the relation of mathematics to physics in terms of embedding a theory T in a mathematical structure M', in the usual set-theoretic sense of there existing an isomorphism between T and a sub-structure M of M'. M' is then taken to be a non-simple conservative extension of M. There is an immediate question regarding the nature of T. To be embedded in M' it must already be 'mathematized' in some form or other. Thus, the issue here is not so much Wigner's inexplicable utility of mathematics in science, in the sense of its being the indispensable language in which theories are expressed, but rather the way in which new theoretical structure can be generated via this embedding of a theory, *which is already mathematized*, into a mathematical structure. Yet this too is an aspect of the unreasonable effectiveness of mathematics and as we shall see, this framework lends itself particularly well to the consideration of the role of group theory in physics.

An uninterpreted calculus C can then be introduced of which T and M can be taken as isomorphic models, and likewise a calculus C' for M' can be presented which drives the introduction of a new theory T' which in turn is partially interpreted via the structure T. This is a useful way of considering the heuristic advantages of re-expressing a theory within a new mathematical structure. In particular, 'surplus' mathematical structure may be heuristically fruitful with respect to further scientific developments. This 'surplus structure' can be represented as the relative complement of T in T', suitably axiomatized (*ibid.*), although it might be more appropriately identified as the extension of the (sub-) structure of the mathematical representation of T' which is isomorphic to T. This meshes nicely with the central idea of T being *embedded* into the 'larger' structure T'. However, the mathematical surplus which, it is claimed, drives these heuristic developments cannot be captured by the usual model-theoretic notion of a structure extension since this simply involves the addition of new elements to the relevant domain with the concomitant new relations. What is required is 'new' structure which is genuinely 'surplus' and which supplies new mathematical resources. Thus an appropriate characterization would be one in which T' is itself embedded in an entire family of further structures – T'', T''' and so on – and this may then capture the heuristic role of mathematics in theory construction.[3]

From the realist perspective, aspects of this structure may cease to be regarded as surplus as the relevant terms come to be accorded ontological

[3] This point arose in discussions with Otávio Bueno and is developed in Bueno & French (forthcoming).

reference. However, in some cases, as Redhead notes, there may be no possibility of ontological reference; nevertheless, the surplus structure may play an essential role in the development of a theory. A particularly well-known example is that of the non-symmetric sub-spaces of the Hilbert space for a collection of quantum particles generated by the action of the permutation group. It is this characteristic of these sub-spaces – namely that they are, or are representative of, surplus structure – that is crucial to Redhead and Teller's argument against 'primitive thisness' or 'Transcendental Individuality' in discussions of individuality and quantum mechanics (see Redhead & Teller 1991 and 1992). The historical origins of this example will be considered below.

The model-theoretic approach is then foreshadowed by the decision to identify T with M and T' with M', in order to 'effectively reduce the problem of intertheory relations to one of relations between mathematical structures' (Redhead 1975, p. 89).[4] In his discussion of 'A Theory of Theories', Redhead considers, and dismisses, various methods for representing a 'general theory' in physics, where a 'general' theory is taken to be essentially reducible to n-tuples of numbers. Here he follows Wigner in taking the purpose of laws to be that of establishing correlations between events (Wigner 1972). An event is represented by the value, $\{\psi(\xi)\}$, of a function ψ, which is a many-one mapping from the domain $\{\xi\}$ to the range $\{\psi(\xi)\}$. The function space of all functions from $\{\xi\}$ into $\{\psi(\xi)\}$ is denoted by $\Gamma = \{\psi\}$ (*ibid.*, p. 91). The structure of the sets $\{\xi\}$ and $\{\psi(\xi)\}$ is governed by 'constitutive laws', such as the commutation relations of quantum mechanics. 'Correlative laws', on the other hand, correlate different events and examples include Maxwell's equations and Newton's laws. The relations embodied in such laws get represented by a unary relation defined on Γ, so that each allowed set of concurrent values of the relevant quantities is represented by a point in Γ. The totality of these points is the 'solution space' γ associated with a particular correlative law. If a particular $\psi(\xi)$ is represented by a point in Γ, then a theory can be identified (Redhead's term) with the appropriate γ.[5]

Redhead puts this framework to good use in discussing the difference between mathematical and physical symmetries, their heuristic power and

[4] Strictly speaking this reduction is unnecessary as we already have structures on both sides of the identity.

[5] In his (1980) paper Redhead asserts that on his account theories are *identified* with sets of functions (p. 151, n 1). Strictly speaking this is incompatible with the claim that a 'general theory' is reducible to n-tuples of numbers, since sets of functions cannot be so identified. I am grateful to Otávio Bueno for pointing this out to me.

the relations between the symmetries of different theories. In a subsequent work (Redhead 1980) this approach is employed to analyse the nature and role of models in physics and approximations in particular can be shown to be accommodated in a uniform and straightforward manner.

The obvious point of comparison within the model-theoretic approach is with van Fraassen's state space approach (1980; 1989), particularly with regard to the corresponding role of constitutive and correlative laws in the former and laws of co-existence and succession in the latter (see Bueno & French forthcoming). Both of these accounts can be related to the more general and, perhaps, more fundamental set-theoretic approach advocated by Suppes: 'By defining set-theoretic predicates . . . one can specify either a state space, [a function space], a relational system, or some other representing mathematical structure or class of structures' (Suppe 1989, p. 4). There are two further elaborations which can then be imported from this Suppesian set-theoretic line.

First of all, theories are open to further development and elaboration, where this is driven both by broadly theoretical concerns as well as by the impact of further experimental developments. Theories typically only partially model their respective domains. In order to capture this sense of openness and partiality, so-called 'partial structures' have been introduced into the model theoretic account (Mikenberg et al. 1986; da Costa & French 1990). Such structures (of first-order) can be represented in the following form:

$$M = < A, R_i >$$

where A is a non-empty set, R_i, is a family of (partial) relations, which are partial in the sense that any relation R_i of arity n_i is not necessarily defined for all n_i-tuples of elements of A; (furthermore, in the general case partial functions could also be included in a partial structure). In general, A denotes the set of individuals of the domain of knowledge modelled in the particular case considered, and the family of partial relations R_i models the various relationships which hold among these individuals. Thus, if we write a n-place partial relation R as the triple $< R_1, R_2, R_3 >$ where R_1, R_2, and R_3 are mutually disjoint, thenR_1 can be regarded as the set of n-tuples that belong to R, R_2 the set that does not belong to R and R_3 the set of n-tuples for which it is not known whether they belong to R or not. Partial structures can be introduced into the function space approach in one of two ways: (i) by considering the constitutive functions themselves as partial, in a formal sense; or (ii) by taking the associated relations (Redhead 1975, p. 92) as partial relations in the sense noted above. In both cases the incompleteness in our empirical information can be accommodated in terms of 'partial function spaces' (see Bueno & French forthcoming). Roughly speaking, of course, the function space ceases to be closed off but may open out in a multitude of heuristic directions.

The second issue concerns the relationship between theories and the phenomena (however construed). Within the model-theoretic approach, van Fraassen has adopted Giere's schema according to which the 'theoretical definition' comprises the specification of an appropriate family of models, while the 'theoretical hypothesis' asserts some sort of correspondence (similarity in Giere's case) between the system of interest and one of these models (Giere 1988; van Fraassen 1989). However, there is typically a wide variety and range of models inserted between the phenomena and the most theoretical structures of a theory. Hughes has attempted to extend Giere's framework by rewriting the theoretical hypothesis as the claim that a part of the world as described in terms of one of these 'intermediate' models of the phenomena can be regarded as a system of the kind specified in the theoretical definition (Hughes 1999). This model-theoretic account has been criticized by Cartwright et al. (1995) on the grounds that it cannot capture the nature of model building in physics since much of this proceeds at the 'phenomenological' level which can be clearly delineated from the 'theoretical' (see also Suárez forthcoming). Elsewhere Ladyman and I have argued that this is not so (1997 and forthcoming). We have suggested that the relationship between these models in the hierarchy, from the 'data models' at the bottom, to the highly abstract theoretical models at the top, can be represented in terms of 'partial isomorphisms' holding between the structures concerned. If we have two partial structures

$$M = < A, R_k >$$

and

$$M' = < A', R'_k >$$

(where R_k and R'_k are partial relations as above) then a function f from A to A' is defined to be a partial isomorphism between M and M' if (i) f is a bijective and (ii) for all x and y in A, R_1xy iff $R'_1f(x)f(y)$ and R_2xy iff $R'_2f(x)f(y)$[6]. Of course, if $R_3 = R'_3 = \emptyset$, so that we no longer have partial structures but 'total' ones, then the standard notion of isomorphism can be recovered. This gives a convenient way of representing the nature of idealization in science and has been extended to give a constructive empiricist account of theory change by Bueno (1997).

[6] These developments are not confined to binary relations of course.

Taking this over to the function space approach, what we would have is a nested series of partial function spaces related in this manner. At one end would be those spaces representing the correlations between the elements of the empirical data, corresponding to Suppes's data models. As we move from what might be called the 'phenomenological' to the 'theoretical', further theoretical elements are introduced, with the corresponding structures related by partial isomorphisms as above. This gives a more accurate representation of the complexities of the relationships between theories and evidence in this context (see Bueno & French forthcoming). More importantly in the present case, this weakening of the embedding relationship between structures can be taken over to the relationship between mathematics and science. As indicated above, the crucial element is the embedding of theoretical structures into a particular mathematical structure and then transferring, as it were, the structure of the latter into the former. By employing this weaker notion of partial isomorphism we can better accommodate the incompleteness and openness of the heuristic situation. It is precisely such a framework that is required to capture the developing relationship between group theory and quantum mechanics.

3. The role of group theory in quantum physics

The principal results of Wigner's influential work, *Group Theory and its Application to the Quantum Mechanics of Atomic Spectra* (Wigner 1959) were first published in 1926 and 1927 (Wigner 1926 and 1927) and Wigner notes in the preface that '[t]he initial stimulus for these articles was given by the investigations of Heisenberg and Dirac on the quantum theory of assemblies of identical particles' (1959, p. vi).[7] These were the papers which established the self-consistent framework of quantum statistics. Following the development of what we now call Bose–Einstein statistics, Fermi presented a theory of the ideal gas which incorporated Pauli's Exclusion Principle and gave a very different form of statistical counting (Fermi 1926). However, it was Heisenberg who explicated the connection between these two forms of statistics and the symmetry characteristics of the relevant states of the particle systems. In his first paper (1926a), he showed that two indistinguishable systems which were weakly coupled always behaved like two oscillators for which there were two sets of non-combining states, symmetric and anti-

[7] Wigner also acknowledges von Laue as perhaps the first of the 'older generation' of physicists to recognize the significance of group theory as '. . . the natural tool with which to obtain a first orientation in problems of quantum mechanics' (1959, p. v).

symmetric. That these two sets of states were not connected then followed from the symmetry of the Hamiltonian of the system under a particle permutation. In his second work (1926b) Heisenberg investigated the concrete example of the two electrons in a helium atom and concluded that only those states whose eigenfunctions are anti-symmetric in their electron coordinates can arise in nature.

Dirac had also read Fermi's paper but subsequently claimed to have forgotten it (1977, p.133) and had not seen Heisenberg's at all, although it is mentioned in a note added in proof to his own 1926 contribution (Dirac 1926). This went further in setting the two forms of statistics in their appropriate theoretical context. He began with what he took to be the fundamental requirement that the theory should not make statements about unobservable quantities. Consequently, he insisted, two states which differ only by the interchange of two particles must be counted as only one. Out of the set of all possible two-particle eigenfunctions only the symmetric and anti-symmetric satisfy the conditions that the eigenfunction should correspond to both of the above states and should be sufficient to give the matrix representing any symmetric function of the particles. These results were then extended to any number of non-interacting particles; and with the anti-symmetric function written in determinantal form, Pauli's Exclusion Principle dropped out quite naturally. Dirac explicitly noted that the theory itself could not determine which form was appropriate but that extra-theoretical considerations had to be appealed to. Thus, the solution with symmetrical eigenfunctions cannot be the correct one for the case of electrons in an atom since it allows any number of electrons to be in the same orbit (*ibid.*, p. 670).

It is this analysis of quantum statistics in terms of the permutation of indistinguishable particles which heuristically motivates the construction of the 'bridge' underpinning the embedding of quantum mechanics into group theory.[8] These papers of Heisenberg's and Dirac's are referred to at the very beginning of Wigner's 1926 paper, where he acknowledges Heisenberg as noting that it is possible to give a partition of states into *n*! groups in such a way that an atom in a state in one of the groups cannot pass over into a state of another and that of these groups one can select one in which Pauli's Exclusion Principle holds (corresponding to Fermi–Dirac statistics) and one in which Bose–Einstein statistics holds (Wigner 1926).[9] It is here that Wigner recognizes that the reduction of the state space of a system of

[8] The term 'bridge' is used in a similar context by Pickering (1995); see also French (1997).
[9] I am grateful to Peter Simons for the translation.

indistinguishable particles (such as electrons) into invariant sub-spaces implies a separation of the states of the system into sets of dynamically non-combining states. In the 1927 paper Wigner then applied the analysis of the representations of the three dimensional rotation group to atomic spectra which subsequently formed the basis of his book (Wigner 1927).

Wigner's works are cited by Weyl in his own 1927 paper on group theory and quantum mechanics (Weyl 1927) and throughout his classic work, *The Theory of Groups and Quantum Mechanics* (1931). Later, in 1939, Weyl referred to Wigner's 'leadership' in this regard (specifically with reference to the decomposition into irreducible invariant sub-spaces using Young's symmetry operators; Weyl 1968, vol. III, p. 679). Weyl himself had made profound contributions to group theory, particularly in his three fundamental papers of 1925 and 1926 (Weyl 1925), in which he effectively initiated the study of global Lie groups.[10] Following the work of Schur, Weyl established the complete reducibility of linear representations of semi-simple Lie algebras.[11] In the context of quantum mechanics, this gave a way of deducing the irreducible representations of a particularly important semi-simple Lie group, namely the three-dimensional pure rotation (or orthogonal) group which we shall consider in greater detail below. These representations in particular were familiar to physicists as they are the 'transformation formulae' for vectors, tensors, etc. (Wigner 1959, p. 168).[12] As Weyl noted, it is in the representation of groups by linear transformations that their investigation became a 'connected and complete theory' (1931, p. xxi) and 'it is exactly this mathematically most important part which is necessary for an adequate description of the quantum mechanical relations' (*ibid.*). Furthermore, group theory reveals 'the essential features which are not contingent on a special form of the dynamical laws nor on special assumptions concerning the forces involved' (p. xxi). Indeed, he claims, '*All quantum numbers, with the exception of the so-called principal quantum number, are indices characterizing representations of groups*' (*ibid.*, his emphasis).

Let us look at this relationship between quantum mechanics and group theory more closely: 'An atom or an ion, whose nucleus is considered as a fixed center of force O, possesses two kinds of symmetry properties: (1) the

[10] For a discussion of this work in the context of the history of the relevant mathematics, see Bourbaki (1975), pp. 426–7.

[11] The term 'Lie algebra' was coined by Weyl himself in 1934.

[12] To each rotation s there corresponds a linear unitary operator $U(s)$ induced in the 'system space' by that rotation. The correspondence $U : s \rightarrow U(s)$ which obeys a composition law structurally identical to the composition of rotations is the representation of the rotation group.

laws governing it are spherically symmetric, i.e., invariant under an arbitrary rotation about O; (2) it is invariant under permutation of its f electrons' (Weyl 1968, p. 268). The first kind of symmetry is described by the rotation group, while the second is described by the finite symmetric group of all $f!$ permutations of f things.

The nature of the relationship becomes clear when we focus on the decomposition of Hilbert space into subspaces which are irreducible and invariant with respect to a particular group (Weyl 1968, pp. 275–7). As Wigner realized, such a decomposition corresponds to the separation of the various values which are possible for a physical quantity. Thus the group of rotations in 'actual' space induces a group of transformations in Hilbert space under which the latter decomposes into invariant subspaces. In each such subspace the rotation group has a definite representation. Correspondingly, the angular momentum operator is separated into 'partial' operators (Weyl's term) each acting on one of the subspaces. The different irreducible representations of the rotation group can be distinguished by an index $j = 0, 1/2, 1, 3/2, \ldots$ and the subspace in which the representation with index j is induced has $2j + 1$ dimensions. Hence, as Weyl points out, we know the angular momentum in the subspace 'independently of the dynamical structure of the physical system under consideration' (1968, p. 276) – its components are the operators which correspond to the infinitesimal rotations in the relevant subspace and the corresponding eigenvalues belong to the relevant representations.

Physically, of course, the value of $2j+1$ gives the 'degree of multiplicity' of an energy level.[13] To each such level of an n-electron system there corresponds a representation of the nth degree (Weyl 1968, p. 276). However, not all these representations actually occur, 'for reasons which cannot be explained without a discussion of electron spin and the Pauli principle' (*ibid.*). That is, group theory gives us surplus structure which we rule out for purposes of application through the invocation of further physical principles, themselves to be embedded within the mathematics (as Dirac indicated in the case of the Exclusion Principle).

The power of group theory reveals itself when we consider the spectrum of an n-electron atom, where $n > 1$. In such cases, the appropriate Schrödinger equation cannot be solved exactly because of the mutual repulsion of the electrons (giving a computationally intractable expression for the potential energy). Hence approximation was viewed as the only way forward (Wigner

[13] As Wigner noted, '[t]he concept of a multiplet system . . . is . . . alien to classical theory' (Wigner 1959, p. 182).

1959, p. 180): we ignore the mutual repulsion of the electrons, solve the resulting Schrödinger equation and reintroduce the effects of the electrons on one another as a perturbation. The perturbation partially removes the degeneracy which results from using this initially crude approximation in which a number of eigenfunctions correspond to each of the resultant eigenvalues; in effect, the energy levels split. Nevertheless, most of the resulting levels are still degenerate and about them, 'nothing is known on a purely theoretical basis (apart from a rough estimate of their positions) except their symmetry properties' (*ibid.*, p. 181). It is here that group theory comes in (and strictly speaking to each level there corresponds three representations, one of the rotation group, one of the reflection or inversion group and one of the permutation group to be discussed below).

How then to proceed? Ignoring the indistinguishability of the electrons and taking the rotation group as the sole symmetry group of the Schrödinger equation, the 'vector addition model' was employed, based on the 'building-up principle' (or 'Aufbauprinzip'), in which the angular momenta of the single electrons are added together in order to give the total (Wigner 1959, pp. 184–93 for pitiless details). As Weyl notes,

> The mathematical interpretation given this model by quantum mechanics is characterized by the two circumstances:
> (1) The determination of the various numerical possibilities is to be interpreted as decomposition into invariant irreducible sub-spaces;
> (2) The addition of vectors has its mathematical counterpart in the multiplicative composition of the representations induced in these sub-spaces. (1968, p. 279)

This 'interpretation' accounts for the following results:

(i) that j is restricted to the values 0, 1/2, 1, 3/2, . . .;

(ii) that the square of the absolute value of the angular momentum is $j(j+1)$ (rather than j^2 as in the classical case);

(iii) that the 'inner' quantum number obtained by the 'composition' of two systems is restricted to the values $|j-j'|, |j-j'|+1, \ldots, j+j'-1, j+j'$ (Weyl 1968, p. 280; Wigner 1959, pp. 187ff)

Distinguishing between the spin and orbital quantum numbers one arrives 'naturally', according to Weyl, at Hund's empirically successful vector model of atomic spectra. But, he emphasizes, this model is not to be taken literally since it can now be obtained within this new 'interpretation' according to which the above results are explained, not by *ad hoc* hypotheses, but on the basis of a unified viewpoint. 'This is the service rendered by the new quantum theory' (Weyl 1968, p. 281). And this was one of the results that Wigner initially thought were the most significant in his work.[14]

There are two further points to note about this. The first concerns the role of group theory in supplying a taxonomy by means of which the 'levels zoo' can be classified. This role is well known, particularly when one considers the development of isospin, SU(3) and so on, but it is worth bringing it out here. What was previously a disparate set of spectroscopic rules, explained in an *ad hoc* and unsatisfactory manner comes to be embedded within a unified, coherent framework. The second point concerns the fundamental importance of group theory in producing such results given the computational intractability of the Schrödinger equation for the many electron atom. At the beginning of his preface Wigner writes that 'The actual solution of quantum mechanical equations is, in general, so difficult that one obtains by direct calculation only crude approximations to the real solutions. It is gratifying, therefore, that a large part of the relevant results can be deduced by considering the fundamental symmetry operations' (1959, p. v).[15] What we have, roughly, is the following sort of scheme: idealizations are introduced, such as ignoring the indistinguishable nature of the particles, giving rise to models which can then be structurally embedded in group theory, which in turn is used to generate the appropriate results. There is a great deal more which needs to be said here, particularly with regard to the 'deduction' that Wigner highlights, but the relationships between such idealized models can be appropriately captured in terms of partial isomorphisms (French & Ladyman 1997 and forthcoming).

Let us now consider Weyl's second kind of symmetry property, namely invariance under particle permutations. If we take two individuals, such as electrons in an atom, which are 'fully equivalent', as Weyl puts it, then the appropriate system space is reducible into two independent subspaces – the space of symmetric and anti-symmetric tensors of second order. Physical quantities pertaining to the system then have 'only an objective physical significance if they depend *symmetrically* on the two individuals' (1931, p. 239; Weyl's emphasis). It can then be shown, in particular, that every possible interaction between the individuals depends symmetrically on them. Hence if the system is at any time in a state in one of these subspaces, no

[14] Subsequently, he writes, he came to agree with von Laue that '. . . the recognition that almost all rules of spectroscopy follow from the symmetry of the problem is the most remarkable result' (1959, p. v).

[15] Redhead has also emphasized this aspect of scientific practice in the context of a discussion of models in physics and he explicitly considers the example of the N-particle Schrödinger equation (1980, p. 156).

influence whatsoever can ever take it out of it (*ibid*., p. 240). '[W]e expect Nature to make use of but one of these sub-spaces' (*ibid.*) but the formalism itself gives us no clue as to which one. The particular sub-space in which a system is located is an initial condition, as it were, determined by the kind of particle considered, boson or fermion. As far as the representation of fermion systems is concerned, the symmetric subspaces are just so much surplus structure.

It can further be shown that every invariant subspace of the state space of f equivalent individuals, including the state space itself, can be completely reduced into irreducible invariant subspaces (1931, p. 301, thm 4.11). As Wigner originally recognized, this reduction implies a separation of the states of the physical system into sets 'which no dynamical influence whatever can cause to enter into combination with each other' (Weyl 1931, p. 320). Of course, as Weyl continually insists, as far as physics is concerned it is only the symmetric and anti-symmetric subspaces that are of interest and, further, only the latter when electrons are being considered:

> The various primitive sub-spaces are, so to speak, worlds which are fully isolated from one another. But such a situation is repugnant to Nature, who wishes to relate everything with everything. She has accordingly avoided this distressing situation by annihilating all these possible worlds except one – or better, she has never allowed them to come into existence! The one which she has spared is that one which is represented by anti-symmetric tensors, and this is the content of Pauli's exclusion principle. (1968, p. 288)[16]

The other, 'surplus' subspaces were subsequently shown to correspond to 'mixed-symmetry' state functions describing so-called parastatistics.

On this basis, one obtains not only the group theoretic classification of line spectra of an atom consisting of an arbitrary number of electrons, taking into account the exclusion principle and spin (Weyl 1931, ch. V),[17] but also an understanding of the nature of the homopolar molecular bond (pp. 341–2) and valency in general (pp. 372–7).

Thus, with regard to the construction of the 'bridge' between the theoretical and mathematical structures, represented by T and M' above, on the quantum mechanical side we have the reduction of the state space into

[16] See also Weyl (1931) pp. 238 and 347. cf. Huggett: '. . . there is no mystery at all about why non-symmetric states are never realized; they are not within the symmetrized Hilbert space that correctly represents the world, and hence do not correspond to physical possibility' (1995, p. 74).

[17] Referring to developments in spectroscopy, Weyl writes 'The theory of groups offers the appropriate mathematical tool for the description of the order thus won' (1931, p. 245).

irreducible subspaces and on the group theoretical side we have the reduction of representations. It is here we have the (partial) isomorphism between (partial) structures, (weakly) embedding T into M'. Underpinning all of this is the intimate relation between the representations of the group of all unitary transformations or the group of all homogeneous linear transformations and those of the symmetric group of permutations of f things: 'the substratum of a representation of the former consists of the linear manifold of all tensors of order f which satisfy certain symmetry conditions, and the symmetry properties of a tensor are expressed by linear relations between it and the tensors obtained from it by the $f!$ permutations' (Weyl 1931, p. 281). Elsewhere Weyl himself refers to this correspondence between the representations as the 'bridge' and notes that since continuous groups are easier to handle than discrete, it leads from the character of the unitary group to that of the permutation group (1968, pp. 286–7).

In particular the above reduction of the state space of equivalent particles into irreducible subspaces 'parallels' (Weyl 1931, p. 321) the complete reduction of the total group space of the symmetric permutation group into invariant subspaces. This 'reciprocity' between the symmetric permutation group and the algebra of symmetric transformations is referred to as 'the guiding principle' in Weyl's work (*ibid.*, p. 377) and elsewhere he writes that

> The theory of groups is the appropriate language for the expression of the general qualitative laws which obtain in the atomic world. In particular the reciprocity laws between the representations of the symmetry group σ_ν and the unitary group Γ are the most characteristic feature of the development which I have here indicated; they have not previously come into their own in the physical literature, in spite of the fact that quantum physics leads very naturally to this relation. (1968, p. 291)

Interestingly, then, the construction of this bridge between quantum mechanics and group theory crucially depends on a further one within group theory itself – the bridge that Weyl identified between the representations of the symmetry and unitary groups as expressed in the reciprocity laws. If one is to capture the relation between group theory and quantum mechanics in the manner that Redhead has suggested, it would seem that the 'surplus structure' should be appropriately characterized in terms of a *family* of structures, as indicated above. In this way we can represent not only the relationships between the mathematics and the physics, as it were, but also the 'internal' relationships within group theory, as manifested in these reciprocity laws. Weyl himself contributed to the elucidation of these relationships and, in the reduction of the representations of the orthogonal group, extended the group theoretical side in such a way that spin could subsequently be accommodated. In particular his approach yields the so-called 'double valued representations' (or, more accurately, the representations of a double 'covering' which came to be called the spinor group) which play a crucial role in spin theory (see Wigner 1959, pp. 157–70).

4. Function spaces again

How does all of this look from the perspective of the model-theoretic approach? Essentially what we have is a theory, quantum mechanics, which is already profoundly mathematized – in terms of Hilbert space, state vectors and so on – and which is then embedded in a mathematical structure, namely group theory, in the manner indicated above. Weyl makes this explicit, with his talk of mathematical 'interpretation' and 'counterparts'. It is the isomorphisms between these counterparts in the sub-structure of group theory and elements of quantum mechanics that formally constitutes the embedding relation. In the case of the rotation group, the embedding is characterized by Weyl's two 'circumstances': first, that the determination of the various possible eigenvalues is to be interpreted as a decomposition into invariant, irreducible subspaces; and secondly, that the addition of state vectors has its mathematical counterpart in the multiplicative composition of the representations 'induced' in these subspaces.

There are three important points to note: first of all, heuristically speaking, this is obviously not a situation in which structure is imported from a well confirmed theory modelling one domain into the domain of other phenomena, as with gauge invariance for example (French 1997). Rather, what we have is the establishment of a correspondence between part of physical theory – quantum statistics – and an aspect of mathematics – group theory – which then motivates the embedding of the former into the latter. Secondly, there were clearly enormous advantages to re-expressing quantum mechanics within this new mathematical structure as a disparate set of *ad hoc* models and broadly phenomenological principles were brought together into a coherent framework. Indeed, as far as Weyl was concerned, the result was nothing less than a 'new' quantum theory. Thus we have a nice example of Redhead's schema.

Thirdly, the situation was not a static one, on either side of the 'bridge'. Both group theory and quantum mechanics were evolving at this time[18] and

[18] As Sarah Kattau has pointed out, one shouldn't place too much emphasis on the fortuitous nature of these twin developments since Euler may have come close to formulating the axioms of group theory (Speiser, 1987).

both exhibited a certain structural openness which allowed for such further developments – a particular example being the incorporation of spin on the one hand and the elaboration of spinor theory on the other. Thus the appropriate model-theoretic formulation would be one involving partial structures in general or partial function spaces in particular[19] and the relations between the corresponding structures would consequently be those of partial isomorphism. Furthermore, each theory, group theory and quantum mechanics, is itself structured, in the manner indicated above. With this framework in hand, of partial isomorphisms holding between complex structures, as represented by a suitable generalization of function space, we can begin to get a better grasp on one of the most important and resonant episodes in the history of modern science. Finally, our understanding of such developments may impact on certain philosophical positions, as we shall now see.

5. Structural realism[20]

Within the realist–antirealist debate in the philosophy of science, there has been recent discussion concerning 'structural realism'. This view can be characterized in the following manner:

> Realism has often been attacked on the grounds that there is a significant lack of convergence in the history of theoretical physics which, so the argument runs, is characterized by discontinuity rather than any continuous cumulative progression. But I believe that detailed historical analysis often reveals more continuity than one suspects, at any rate at the level of structure rather than ontology. To see the distinction in a general sort of way, compare asking the question What is a field? with the question What are the mathematical equations governing its behaviour? (Redhead 1995, p. 18)

Hence, despite the radical changes in ontology that may occur through theory change, there is an underlying continuity of structure (cf. Post

[19] Suppes, of course, gives group theory as an example demonstrating the advantages of employing the model-theoretic approach (1957). The present work can be viewed as an extension of this application in line with his remark that 'The set-theoretical definitions of the theory of mechanics, the theory of thermodynamics, the theory of learning, to give three rather disparate examples, are on all fours with the definitions of the purely mathematical theories of groups, rings, fields, etc. From the philosophical standpoint there is no sharp distinction between pure and applied mathematics . . .' (Suppes 1967, pp. 29–30).

[20] I am immensely grateful to James Ladyman for extensive discussions on this topic, which is the subject of his PhD thesis (further details can be found in Ladyman (forthcoming)) and to Otávio Bueno.

1971). Worrall gives the example of Fresnel and the nature of light – this nature effectively changes as we move from a simple wave theory, through Maxwell's equations of electromagnetism to photons and QED. Structurally, however, Fresnel's equations are incorporated into Maxwell's and the mathematical structure is preserved (Worrall 1989).

Two issues immediately arise in this context: First of all, if it is mathematical structure, rather than 'ontology' that is supposed to provide the continuity the realist needs, then the question can be posed – which structure? Consider the story above. From the structural realist viewpoint, are we to focus on the equations of quantum mechanics, such as Schrödinger's equation, or on what Weyl calls the 'appropriate language', namely group theory? In terms of unifying and systematizing power, it is the latter which should attract our attention. Worrall insists that the relevant structure should be that which plays a role in producing novel predictions from the theory. One well-known example is the group theoretic prediction of the Ω- particle; and although the relevant context is elementary particle physics, it is notable that the use of group theoretic techniques in this domain can be attributed to a return to the study of Wigner's work (Mehra 1971, pp. 329–30).

In the context discussed here things are not quite so straightforward. It is surely not the case that Weyl's work on the double representations 'predicted' spin, in Worrall's sense. Spin, like the spectroscopic multiplets, was already represented in terms of an intermediate model. What the group theoretic approach does, as indicated above, is embed such models in the more abstract representation of semi-simple Lie groups which sets quantum theory in a unitary framework.

This suggests that the hypothetico-deductive view of prediction is too simplistic. In such an account statements referring to the predicted phenomena are logically deduced from the statements of the theory which in turn incorporate the relevant mathematics. This fails to capture the complexities of the relationship between theory and phenomena, where the latter may be theoretically represented at an intermediate level. A more sophisticated framework is offered by the model-theoretic approach in which one has a hierarchy of models between the 'higher' theoretical structures and the lower phenomenological and empirical ones. The relationships between these structures can then be appropriately represented by means of partial function spaces and associated partial isomorphisms.

There is also a form of horizontal 'spreading' as models are conjoined in order to explain a given phenomenon. The ability to capture this is one of the advantages of the model-theoretic approach which has been emphasized by Lloyd (1988). A similar situation holds for prediction: consider Dirac's famous prediction of the existence of anti-particles. As is well known, this drops out of his relativistic wave equation; and so according to Worrall, we should be structural realists about the latter. But this equation in turn involved a generalization of the Pauli spin matrices from 2×2 matrices to 4×4, which provided the crucial extra 'degree of freedom' necessary to allow for the possibility of anti-matter. Following the formal connections, why should we remain at the level of Dirac's equation and be realists about *that*? Where do we draw the line between that particular piece of mathematics and other structural elements that come into view as we move both vertically and horizontally through the nexus?

The second issue concerns the ontology itself. Returning to the Redhead quote, notice that the field is still present in our ontology on this view, its nature unknown and unknowable (Worrall 1989). There is, then, a split between the nature of a theoretical entity and its structural representation, a split which Psillos, for example, finds untenable: the 'nature' of a thing can be described entirely in terms of this structural representation and structural realism collapses into the common or garden variety (Psillos 1995). However, this is to ignore the metaphysical 'packages' in terms of which the nature of things is to be understood. Redhead's first question above – 'What is a field?' – can be fully answered only by appealing to such metaphysics and this answer is not exhausted by the answer to the second question – 'What are the mathematical equations governing its behaviour?'

In the case of quantum mechanics, these metaphysical elements are underdetermined by the formalism of the relevant theory itself as the example of individuality in quantum mechanics demonstrates: either the entities are regarded as individuals subject to state accessibility restrictions, or they are regarded as non-individuals in some sense (Redhead & French 1988). It is precisely at this point that the 'standard' realist finds herself unable to answer the parallel question to Redhead's above, namely what is a quantum entity? This partly motivates the move to an 'ontological' form of structural realism, a version of which has been recently defended by Ladyman (1998). Here there are no unknowable *objects* lurking in the shadows and *objectivity* is understood structurally, in terms of the relevant set of invariants.[21] Thus Born, for example, writes: 'Invariants are the concepts of which science speaks in the same way as ordinary language speaks of 'things' and which

[21] As Weyl notes, '. . . the investigation of the invariants of a group is tied up with the ascertainment of its representations' (1968, p. 682).

it provides with names as if they were ordinary things' (Born 1956, p.163).[22] With these invariants understood and represented group theoretically, we arrive at a kind of structural realism which takes structure seriously.[23]

However, it has been argued that it makes no sense to talk of structure without its component elements (Redhead, private discussion): how can we talk of a group if we have done away with the elements which are grouped? A response can be given by focusing on the role of the metaphysical elements in this particular account. We begin with a conceptualization of the phenomena – the flashes on a scintillation screen, say – informed by a broadly classical metaphysics (in the philosophical rather than physical sense) in terms of which the entities involved are categorized as individuals. That categorization is projected into the quantum domain, where it breaks down and the fracture with the classical understanding is driven by the introduction of group theory: the entities are classified via the permutation group which imposes perhaps the most basic division into 'natural kinds', namely bosons and fermions. It is over this bridge that group theory is related to quantum mechanics as indicated above.

However, by insisting that particle permutations are not observables, the theory (of quantum mechanics, as informed group theoretically) leaves us in the situation of metaphysical underdetermination. The theory itself provides no guide to ontology in this sense (where this is understood as having a metaphysical dimension) and we have the interesting situation in which the whole enterprise gets off the ground on the back of a classical metaphysics, which is then effectively discarded. Thus the elements themselves, regarded as individuals, have only a heuristic role in allowing for the introduction of the structures which then carry the ontological weight.

Perhaps the last word should be left to Weyl himself, who writes, regarding the 'essential features' revealed by group theory, 'We may well expect that it is just this part of quantum physics which is most certain of a lasting place' (1931, p. xxi).

References

Born, M. (1956). *Physics in My Generation* (London: Pergamon Press).

Bourbaki, N. (1975). *Lie Groups and Lie Algebras*, part I (Paris: Hermann).

[22] Gingerich (1975), pp. 558–9, cites similar sentiments by Heisenberg.

[23] This suggestion was presented at a Sigma Club conference at Cambridge in May 1995 and is developed in Ladyman & French (forthcoming).

Bueno, O. (1997). 'Empirical Adequacy: A Partial Structures Approach', *Studies in History and Philosophy of Science* **28**, 585–610.

Bueno, O. & S. French (forthcoming). 'State Spaces, Quasi-Truth and the Semantic Approach', preprint.

Cartwright, N., T. Shomar, & M. Suárez (1996). 'The Tool Box of Science: Tools for Building of Models with a Superconductivity Example', in W. E. Herfel, W. Krajewski, I. Niiniluoto & R. Wójcicki (eds.), *Theories and Models in Science*, Poznán Studies in the Philosophy of the Sciences and the Humanities (Amsterdam: Rodopi) pp. 137–49.

Da Costa, N. C. A. & S. French (1990). 'The Model-Theoretic Approach in the Philosophy of Science', *Philosophy of Science* **57**, 248–65.

Dirac, P. A. M. (1926). 'On the Theory of Quantum Mechanics', *Proceedings of the Royal Society A* **112**, 661–77.

(1977). 'Recollections of an Exciting Era', in C. Weiner (ed.), *History of Twentieth Century Physics*, Proceedings of the International School of Physics 'Enrico Fermi', Course LVII (New York and London: Academic Press) pp. 109–46.

Doncel, M. A. et al. (1987). *Symmetries in Physics (1600–1980)*, Proceedings of the 1st International Meeting of the History of Scientific Ideas (Barcelona: Bellaterra).

Fermi, E. (1926). 'Zur Quantelung des idealen einatomigen Gases', *Zeitschrift für Physik* **36**, 902–12.

French, S. (1997). 'Partiality, Pursuit and Practice', in M. L. Dalla Chiara et al., *Proceedings of the 10th International Congress on Logic, Methodology and Philosophy of Science* (Dordrecht: Kluwer), pp. 35–52.

French, S. and Ladyman, J. (forthcoming). 'Reinflating the Semantic Approach', *International Studies in the Philosophy of Science*.

(1997). 'Superconductivity and Structures: Revisiting the London Account', *Studies in History and Philosophy of Modern Physics* **28**, 363–93.

French, S. & M. Redhead (1988). 'Quantum physics and the identity of indiscernibles', *British Journal for the Philosophy of Science* **39**, 233–46.

Giere, R. N. (1988). *Explaining Science* (Chicago: University of Chicago Press).

Gingerich, O. (1975). *The Nature of Scientific Discovery* (Washington: Smithsonian Institution Press).

Heisenberg, W. (1926a). 'Mehrkörperproblem und Resonanz in der Quantenmechanik', *Zeitschrift für Physik* **38**, 411–26.

(1926b). 'Über die Spektra von Atomsystemen mit zwei Elektronen', *Zeitschrift für Physik* **39**, 499–518.

Huggett, N. (1995). 'What are Quanta, and Why Does it Matter?', in D. Hull et al. (eds.), *PSA 1994*, Vol. 2 (East Lansing, MI: Philosophy of Science Association, 1994), pp. 69–76.

Hughes, R. I. G. (forthcoming). 'Models, Brownian Motion and the Disunities of Physics', in John Earman and John D. Norton (eds.), *The Cosmos of Science* (Pittsburgh: University of Pittsburgh Press).

Ladyman, J. (1998). 'What Is Structural Realism?, *Studies in History and Philosophy of Science*, **29**, pp. 409–24.

Ladyman, J. & S. French (forthcoming). 'The Prospects for Structural Realism', preprint.

Lloyd, E. (1988). *The Structure and Confirmation of Evolutionary Theory* (Princeton: Greenwood Press).

Mehra, J. (1971). *The Solvay Conferences in Physics* (Dordrecht: Reidel).

Mikenberg, I., N. C. A. da Costa & R. Chuaqui (1986). 'Pragmatic Truth and Approximation to Truth', *Journal of Symbolic Logic* **51**, 201–21.

Pagonis, C. (1996). 'Quantum Mechanics and Scientific Realism', Cambridge University: PhD thesis, Cambridge University.

Pickering, A. (1995). *The Mangle of Practice* (Chicago: University of Chicago Press).

Post, H. R. (1971). 'Correspondence, Invariance and Heuristics', *Studies in History and Philosophy of Science* **2**, 213–55.

Psillos, S. (1995). 'Is Structural Realism the Best of Both Worlds?', *Dialectica* **49**, 15–46.

Redhead, M. L. G. (1975). 'Symmetry in Intertheory Relations', *Synthese* **32**, 77–112.

(1980). 'Models in Physics', *British Journal for the Philosophy of Science* **31**, 145–63.

(1995). *From Physics to Metaphysics* (Cambridge: Cambridge University Press).

Redhead, M. & P. Teller (1991). 'Particles, Particle Labels, and Quanta: The Toll of Unacknowledged Metaphysics', *Foundations of Physics* **21**, 43–62.

(1992). 'Particle Labels and the Theory of Indistinguishable Particles in Quantum Mechanics', *British Journal for the Philosophy of Science* **43**, 201–18.

Speiser, D. (1987). 'The Principle of Relativity in Euler's Work', in Doncel et al. (1987), pp. 31–47.

Suárez, M. (1999). 'The Role of Models in the Application of Scientific Theories: Epistemological Implications', in M. Morgan and M. Morrison (eds.), *Models as Mediating Instruments* (Cambridge: Cambridge University Press).

Suppe, F. (1989). *The Semantic Conception of Theories and Scientific Realism* (Urbana, IL: University of Illinois Press).

Suppes, P. (1957). *Introduction to Logic* (New York: Van Nostrand).

(1967). *Set-Theoretical Structures in Science*; mimeograph, Stanford University.

Van Fraassen, B. (1980). *The Scientific Image* (Oxford: Clarendon Press).

(1989). *Laws and Symmetry* (Oxford: Clarendon Press).

Weyl, H. (1925). 'Theorie der Darstellung kontinuierlicher halb-einfacher Gruppen durch lineare Transformationen, I', *Mathematische Zeitschrift* **23**, 271–309; II, **24** (1926) 328–76, III, **24** (1926) pp. 377–95; reprinted in Weyl (1968), vol. II, pp. 543–647.

(1927). 'Quantenmechanik und Gruppentheorie', *Zeitschrift für Physik* **46**, 1–46; also in Weyl (1968), vol. III, pp. 90–135.

(1928). *The Theory of Groups and Quantum Mechanics* (London: Methuen and Co.; English translation, 1931).

(1968). *Gesammelte Abhandlungen* (Berlin: Springer-Verlag).

Wigner, E. P. (1926). 'Über nicht kombinierende Terme in der neueren Quantentheorie', *Zeitschrift für Physik* **40**, 492–500 and 883–92.

(1927). 'Einige Folgerungen aus der Schrödingerschen Theories für die Termstrukturen', *Zeitschrift für Physik* **43**, 624–57.

(1931). *Group Theory and Its Application to the Quantum Mechanics of Atomic Spectra* (London: Academic Press; English translation, 1959).

(1972). 'Events, Laws of Nature and Invariance Principles', in *Nobel Lectures; Physics 1963–70* (Amsterdam: Elsevier), pp. 6–17.

Worrall, J. (1989). 'Structural Realism: The Best of Both Worlds', *Dialectica* **43**, 99–124.

9

Can the fundamental laws of nature be the results of evolution?

ABNER SHIMONY

1. Introduction

The title of my essay, 'Can the fundamental laws of nature be the results of evolution?', may suggest to some people that I shall discuss the history of science. This is not my intention. We have attained our present state of knowledge (presumably not the final state) of fundamental laws by an intricate historical process, and it is interesting to inquire how accurately, and with what reservations, the term 'evolution' in the Darwinian sense applies to the historical process of scientific development. (Tangentially, I remark that I have strong reservations about characterizing the development of the natural sciences as 'evolutionary', since that development is driven by a final cause – to find out the truth about nature – while the elimination of teleology is central to Darwinism.) My intention, however, is to discuss the laws themselves, as matters of fact concerning nature, rather than the knowledge of these laws by human beings. I wish to honour Michael Redhead by sharing his priorities: giving precedence to questions about the constitution of the world over questions about human knowledge.

But many people are baffled by this clarification. They willingly admit that secondary or derivative laws of nature may be the products of evolution, but cannot see how the basic laws themselves can evolve. Thus, the genetic code is certainly a law of biology, but it is presumably the result of billions of years of trial and error on the part of interacting molecular species; such, at any rate, is the thesis of prebiotic evolution. But the basic laws of physics are usually assumed to provide a constant background for the trial and error of natural selection that took place while precursors of the present genetic code were explored and assessed. It would make no sense, accordingly, to regard the basic laws of nature as products of evolution, because there would be nothing to provide the requisite constant background for the trial and error that produced them. Why, then, should we pay any attention to a manifestly irresponsible extension of the concept of evolution beyond its proper domain?

There are, I think, good reasons for examining without prejudice the conjecture that the basic laws of nature of products of evolution. One reason is respect for the depth and ingenuity of its proponents, which include the philosophers Charles Saunders Peirce and Alfred North Whitehead and the philosophically inclined theoretical physicists John Archibald Wheeler and Lee Smolin. A second reason is the difficulty of understanding the character of the modality *necessity* when that modality is applied to basic natural laws. By contrast, there have been some well articulated theories of the necessity of mathematical propositions (the best, in my opinion, being the platonism of Gödel, summarized in Wang (1986, pp.13–15)). After one has tried strenuously to understand in exactly what sense basic natural laws are necessary, and has become discouraged with the answers offered by various kinds of rationalism, conventionalism, Kantianism, etc., one feels in desperation the appeal of the evolutionary conjecture. A third reason is that the evolutionary conjecture leads one to explore a fairly large number of deep and fascinating philosophical and physical ideas, and even if at present we cannot assess these ideas definitively, we can derive great intellectual stimulation from thinking about them. In other words, there is something festive about the matters that I propose to discuss, and after all, festivity is appropriate in this volume.

The question I am addressing, on the status of fundamental laws of nature, belongs to a family of metaphysical questions, among them 'What is the ontological status of a fundamental law of nature?', 'If a fundamental law of nature is necessary, what is the character of its necessity, as contrasted, for example, to the necessity of a true proposition of number theory?', and 'What kind of reason for being *(ratio essendi)* can properly be offered for a fundamental law of nature, as contrasted with a reason for asserting its truth (*ratio cognoscendi*)?' This family of questions presupposes that there are such things as natural laws and that they are facts concerning the world, so that asking about their ontological status is not an empty exercise. This presupposition would presumably be denied by an instrumentalist philosopher of science, who regards statements of laws in scientific discourse as no more than devices for unifying and systematizing the data of experience. I shall not examine the instrumentalist–realist controversy here, but shall refer to Redhead's excellent case against instrumentalism and for realism (1995, chap. 1).

The best candidates we have at present for fundamental natural laws are the principles of quantum mechanics, the principles of general relativity, and the standard model of elementary particle theory. It is likely that none of these is literally true. There are considerations that indicate the need to go beyond each of these theories, for instance the difficulty of combining general relativity and quantum theory. Nevertheless, the predictive and unificatory power of each of these three theories is so great that almost certainly they are 'good approximations', in some appropriate sense, to the better theories that eventually will replace them. And each of them is non-derivative in our present state of physical knowledge, which makes them suitable in our investigation of the ontological status of fundamental laws. Hence, even as we acknowledge the likelihood that they will be (honourably) displaced from the scientific world view, we can use them as respectable surrogates for the as-yet-unknown replacements.

2. Leibniz's principle of sufficient reason

An appropriate historical starting-point for analysing the status of fundamental laws is the metaphysics and natural philosophy of Leibniz. He treats the question rationalistically, but does so with (it seems to me) an unprecedented clarity regarding the difference between mathematical and physical truth. And his rationalistic treatment deploys ideas, such as 'all possible worlds', 'actual world', and 'sufficient reason', which figure even in later treatments that depart from Leibniz's rationalism. Here are two relevant passages:

> There are two first principles of all reasonings, the principle of contradiction . . . and the principle that a reason must be given, i.e., that every true proposition, which is not known *per se*, has an *a priori* proof, or that reason can be given for every truth, or as is commonly said, that nothing happens without a cause. Arithmetic and Geometry do not need this principle, but Physics and Mechanics do. The true cause, why certain things exist rather than others, is to be derived from the free decree of the divine will, the first of which is to will to do all things in the best way. (1875 G.VII.309, quoted by Russell 1900, pp. 209–10)

> As there are an infinity of possible worlds, there are also an infinity of laws, some proper to one, others to another, and each possible individual of any world contains in its notion the laws of its world. (1875 G.VII.40, quoted by Russell 1900, p. 210)

Leibniz is too sanguine in his philosophy of mathematics about the possibility of actually deducing logical and mathematical truths from the principle of contradiction. Even the true propositions of predicate logic require a richer basis than Leibniz realized, and *a fortiori* regarding the true propositions of number theory and set theory. Wang (1986, p.14 and in many other loci) cites and approves Gödel's emendation that a mathematical truth is analytic in the sense that it holds 'owing to the meaning of the concepts occurring in it'. The emendation does not, however, weaken Leibniz's distinction between mathematical truths and the laws of physics, the former holding in all possible worlds and the latter only in some of them.

How does Leibniz apply the principle of sufficient reason? In the case of a derivative law of physics the answer is straightforward: the sufficient reason consists in more basic laws together with boundary conditions and other relevant singular facts. What is distinctive about the metaphysics within which Leibniz's natural philosophy is embedded is the resort to theology in order to supply a sufficient reason for a fundamental law. He postulates that the conjunction of God's desire to choose the best possible world and his power to execute this choice constitutes the requisite sufficient reason.

Leibniz's invocation of theology to preserve the principle of sufficient reason when one arrives at the extreme limit of physical law raises many further questions, the crucial one being whether the principle of sufficient reason holds in theology itself. Leibniz answers this question positively, thereby completing the structure of his rationalist philosophy, by accepting and refining Anselm's ontological argument. He argues for the possibility of a most perfect being and then reasons that a most perfect being (i.e., God) necessarily exists, because existence is part of the concept of perfection (G.VII,261, Russell 1900, p. 287). Now one can admire the design of Leibniz's philosophical architecture and still be sceptical of its robustness. My own scepticism derives from the penetrating argument of Kant (1781, pp. 592–602) that existence is not a predicate and hence cannot be established by unpacking the concept of perfection. But Leibniz's admirable philosophical architecture then has the devastating consequence that the applicability of the principle of sufficient reason to fundamental laws of physics is as vulnerable as the ontological argument for the existence of God, and therefore very vulnerable indeed.

A fresh start is needed on the question of the ontological status of fundamental natural laws, a start which abandons the ambition of extreme rationalism. Some advice about strategy can be learned from the German entomologist Stein in Joseph Conrad's *Lord Jim*. Stein says,

> 'A man that is born falls into a dream like a man who falls into the sea. If he tries to climb out into the air as inexperienced people endeavour to do, he drowns – nicht wahr? . . . No! I tell you! The way is to the destructive element submit yourself, and with the exertions of your hands and feet in the water make the deep, deep sea keep you up.' (Conrad 1900, ch. 20)

The advocates of an evolutionary explanation of fundamental natural laws are the followers of Stein, and the 'destructive element' in which they seek support is the sea of contingency.

3. The theory of evolution

What inspired a few scientists and philosophers to propose that fundamental natural law emerges out of contingency is the intellectual coherence of the Darwinian, and later the neo-Darwinian, theories of biological evolution. Before examining the evolutionary proposals of Peirce, Whitehead, Smolin, and Wheeler, it will be useful to summarize the structure of the modern version of the theory of evolution.[1]

The theory has three major components: a theory of variation and mutation, a theory of selection, and a theory of inheritance (explaining how the mutations that are selected can be preserved). Of these, the first and third have developed radically in the last century and a half: from Darwin to Mendel to Morgan to Watson and Crick to contemporary biochemistry the theories of variation and inheritance have been wonderfully refined. I maintain, by contrast, that no new fundamental principles have been brought into the theory of natural selection since Darwin's time, for the simple reason that the theory of natural selection never did have any principles of its own and does not need any. I have argued elsewhere (1989) that there are undoubtedly phenomena of natural selection and there is a theory of natural selection, but it is a theory without principles of its own. Essentially, the theory of natural selection is nothing more than the systematic deployment of information about organisms and their environment for the purpose of evaluating (quantitively if possible, otherwise at least comparatively) the probabilities of survival and reproduction of variant organic lineages. The considerations which determine these probabilities are drawn from other branches of biology and from the physical sciences of the environment. There surely has been progress since Darwin's time in the theory of

[1] My summary draws on such wide and unsystematic reading that I cannot give sources for individual theses, but Monod (1972) was the source that I found most inspiring.

natural selection, because of increased sophistication concerning strategies and statistics of biological survival. But there was no refinement of the principle of natural selection, because there never was such a principle, neither in Darwin's work nor thereafter. When one examines the propositions which have been set forth as proposed principles of natural selections, one always finds that they are either vacuous or derivative from other bodies of biological theory.

I stress the thesis of the non-existence of a principle of natural selection, not just because I claim some credit for articulating it, but because it is indispensable to the advocates of an evolutionary explanation of fundamental laws. One obviously would not achieve 'law without law', as Wheeler (1983) has named this programme, if it depended upon a principle of natural selection which itself was a fundamental law. Indeed, it seems to me that the advocates of an evolutionary explanation of fundamental law assume the non-existence of a principle of natural selection, even though they do not say so explicitly.

4. Peirce

The pioneer of the evolutionary explanation of fundamental laws was Charles Sanders Peirce. He was inspired both by Darwin's theory of natural selection and by the kinetic theory of gases. He was one of the earliest to recognize an underlying similarity of logic of those two theories, especially regarding the emergence of uniformities from the play of chance (Peirce 1877, reprinted in Peirce 1934, paragraph 33). In Peirce's speculative cosmology he proposed a vast extrapolation of this idea of emergence:

> in the beginning – infinitely remote – there was a chaos of unpersonalized feeling, which being without connection or regularity would properly be without existence. This feeling, sporting here and there in pure arbitrariness, would have started the germ of a generalizing tendency. Its other sportings would be evanescent, but this would have a growing tendency. Thus, the tendency to habit would be started; and from this, with the other principles of evolution, all the regularities of the universe would be evolved. (Peirce 1932, paragraph 33)

When I read this passage fifty years ago, as an undergraduate, I found it inspiring. And I still do, especially since I can see how Peirce anticipated by nearly 100 years such ideas as 'order out of disorder', 'strange attractors', and 'law without law'. But as an undergraduate I thought that I understood the passage, whereas now I know that I did not then and do not now.

I believe, however, that I can now identify what is baffling in Peirce's statement.

As pointed out earlier, the theory of evolution has three components – theories of variation, of selection, and of inheritance. Since Peirce gives only the broadest sketch of his idea of cosmic evolution, we cannot expect from him a detailed scenario of variation, selection, and inheritance. But with regard to variation there is a profound conceptual difficulty. In all games of chance the random processes occur in a theatre with a fixed structure, sometimes called the 'Spielraum'. Such is the case also in the great games of chance envisaged by the theories of biological and prebiotic evolution. The theatre of variation according to neo-Darwinian evolutionary theory is provided by the biochemistry of nucleic acids: the exact nature of these molecules and of the chemical forces among them is fixed, and variation consists in novel combinations of the fixed building blocks. The theatre of variation in prebiotic evolution is much larger, constituted by the prebiotic molecules available for combinations and associations. But what conceivably could the theatre be for Peircean cosmic variation? It is doubtful that the range of possibilities within which 'sporting' occurs is well defined unless there is a theatre with some definite structure. How, moreover, can the various 'sportings' be compared with respect to evanescence and stability without some criterion of proximity, which would depend upon a primitive topology in the theatre? If it is allowed, however, that the theatre of cosmic evolution is endowed with some fixed structure, even a very weak one, then there is a retrenchment from the programme of giving an evolutionary account of *every* general law.

5. Whitehead

Just such a retrenchment is exhibited in Alfred North Whitehead's philosophy of organism. Whitehead's metaphysics (1929, 1933), like Leibniz's monadology (1714), supposes the ultimate actual entities of the world to be essentially mental (or proto-mental). Actual entities and societies of them may be roughly characterized by physical attributes, but Whitehead like Leibniz considers these attributes to be abstractions from the rich inner lives of the actual entities. Whitehead differs from Leibniz, however, by ascribing to his actual entities finite – indeed, very brief – temporal duration. And he rejects entirely Leibniz's doctrine that 'the monads are windowless'. Instead, Whitehead postulates that each actual entity, in the initial stage of its genesis, internalizes to some extent the feelings of earlier entities. He introduces a technical term 'prehension', which suggests a relation more primitive than 'apprehension' or 'comprehension', to characterize the recapitulation of an old entity by a new one. Prehension is not deterministic, however, and innovations can occur in the formation of a new entity. These innovations constitute the component of variation in a Whiteheadian theory of cosmic evolution. For the most part, innovations are infinitesimal, but they can have a grand cumulative effect. A kind of natural selection eventuates in the formation of societies with robust characteristics, the society being self-perpetuating in virtue of the prehension of characteristics of the previous members of the society by fresh members. An individual electron and the personality of an individual human being are examples of self-perpetuating societies of actual entities. The vastest species of society is a 'cosmic epoch' (1929, ch. III, section II):

> Thus a system of 'laws' determining reproduction in some portion of the universe gradually rises into dominance; it has its stage of endurance, and passes out of existence with the decay of the society from which it emanates . . . Maxwell's equations of the electromagnetic field hold sway by reason of the throngs of electrons and of protons. Also each electron is a society of electronic occasions, and each proton is a society of protonic occasions. These occasions are the reasons for the electromagnetic laws; but their capacity for reproduction, whereby each electron and each proton has a long life, and whereby new electrons and new protons come into being, is itself due to these same laws. But there is disorder in the sense that the laws are not perfectly obeyed, and that the reproduction is mingled with instances of failure. There is accordingly a transition to new types of order . . . (1929, pp. 139–40)

From this passage one sees that Whitehead regards the fundamental laws of physics as products of evolution, but it is an evolution that takes place in an infinitely vast theatre of potentiality, a theatre governed by general principles which are applicable without exception (Whitehead 1929, p. 138). The proto-mentality of all actual entities, their transience, their prehension of predecessors, their innovations, the integration of the innumerable ingredients of their experience, their 'objective immortality' as ingredients in the experience of subsequent entities – such principles are general but are not the results of evolution. So far as I know, Whitehead does not attempt to provide 'sufficient reasons' for these metaphysical principles. Whitehead like Peirce is an indeterminist, and *ipso facto* limits the domain of validity of the principle of sufficient reason, but unlike Peirce he seems willing to permit not only individual contingencies but also certain general propositions to lie outside that domain of validity. The universe governed by these rationally unjustified general metaphysical propositions can be taken as the theatre within which the Whiteheadian evolution of natural laws takes place.

A plausible place to search for a confirmation of the mortality of the basic laws of nature, as Whitehead conjectures, is in the most remote galaxies our radio telescopes can detect. But the best evidence we now have is non-confirming. The absorption spectra from galaxies fifteen billion light years away are the same as the absorption spectra obtained in terrestrial laboratories from light passing through vapours of known chemical elements, once correction is made for the large Hubble red shift. This agreement would be an astonishing coincidence if terrestrial chemical elements were not present in the remote galaxies and if the laws of atomic physics and electromagnetism on earth were not valid in the remote galaxies. Accordingly, if the evolutionary conjecture is to be confirmed, it may be necessary to look *beyond* the remote galaxies – or equivalently, 'before' the Big Bang or 'after' the Big Crunch (whatever those prepositions and substantives may mean). And that is the proposal of Lee Smolin's book *The Life of the Cosmos* (1997), to which I turn now.

6. Smolin

The standard model of elementary particles requires nineteen fundamental dimensionless parameters, whose role is primarily to determine the relative masses of the various elementary particles and the characteristics of the four fundamental forces (gravitational, weak, electromagnetic, and strong). This proliferation of apparent arbitrariness at the fundamental level is generally regarded as a serious blemish of the standard model, and the removal of this blemish motivates string theory and other speculative elementary particle theories. What Smolin proposes is to keep the standard model but to supplement it with an evolutionary explanation of the values of the nineteen parameters. He makes two postulates:

> The first of these is . . . that quantum effects prevent the formation of singularities, at which time starts or stops . . . then time does not end in the centers of black holes, but continues into some new region of space-time, connected to our universe only in its first moment. Going back towards the alleged first moment of our universe, we find also that our Big Bang could just be the result of such a bounce in a black hole that formed in some other region of space and time . . . If we accept this then we have not only the inevitable inaccessible regions, we have the possibility

> that these regions could be universes as large and as varied as the universe we can see. Moreover, as our own visible universe contains an enormous number of black holes, there must be enormous numbers of these other universes. There are at least as many as there are black holes in our universe, but surely if we can believe this we must believe there are many more than that, for why should not each of these universes also have stars that collapse to black holes and thus spawn new universes. (Smolin 1997, p. 93).

Smolin's second postulate is that

> the basic forms of the laws don't change during the bounce, so that the standard model of particle physics describes the world both before and after the bounce. However, I will assume that the parameters of the standard model do change during the bounce. How do they change? In the absence of any definite information, I will postulate only that these changes are small and random. (Smolin 1997, p. 94).

These two postulates constitute the component of variation in Smolin's theory of cosmic evolution. He proposes a cosmogony in the plural. The variable entities are universes, and the theatre in which the variation occurs is governed by the principles of quantum gravity (as yet not fully constructed) and the form of the standard model.

It is the component of natural selection which I find most striking in Smolin's theory. Universes with different values of the nineteen parameters differ greatly from each other in their propensity to generate other universes. He restricts his attention to universes with sufficient matter that gravitation will slow down and eventually reverse its expansion, leading finally to a collapse and a 'bounce'. Such a universe will produce at least one descendant. But unless the universe produces a star that is sufficiently massive to collapse into a black hole, it will have only one. If we borrow from population genetics the term 'fitness', meaning roughly 'expected number of descendants', then the fitness of a universe is determined by its propensity for producing black holes. The centre of Smolin's argument, which requires the deployment of much astrophysical information plus some extrapolation and conjecture, is the sketch of a demonstration that the range of the nineteen parameters which determine universes with high fitness is extremely narrow. Consequently, in the ensemble of all universes, the distribution of the values of the parameters is highly peaked. Most actual universes have parameter values in the narrow range which generates many descendants. If one then assumes – without a principle of natural selection, just by general probabilistic reasoning – that our universe is with overwhelming probability 'typical', then an explanation is provided for the values of the parameters of the standard model found in *this* universe.

Smolin's theory of cosmic evolution is related to, but much stronger than, the *weak* version of the *Anthropic Principle,* which asserts that 'all physical and cosmological quantities . . . take on values restricted by the requirement that there exist sites where carbon-based life can evolve and the requirement that the universe be old enough for it to have already done so' (Barrow & Tipler 1986, p.16). This version of the Anthropic Principle is indeed weak, because it only makes explicit some necessary conditions for the physical and biological situation with which we are confronted in the actual world. Smolin has a much stronger thesis. First, if the reasoning summarized so far is correct, the ensemble of possible universes has a distribution of values of the parameters of the standard model that is sharply peaked. He then argues further (1997, chs 11–5) that the values of the parameters that are conducive to the production of black holes are also conducive to life. As a consequence life is *typical* in the ensemble of possible universes. It should be remarked, however, that Smolin expresses some reservations about the second part of his argument and acknowledges some scenarios in which life is an atypical phenomenon (1997, p. 321).

Smolin is an enthusiastic Leibnizean, who cites the principle of sufficient reason with approbation from time to time. But even if his programme of cosmic evolution withstands all objections, it is far from providing sufficient reasons for the fundamental laws of physics. Smolin postulates the validity of the formal aspects of the standard model of elementary particles and relies upon evolution only for the purpose of supplying the 'arbitrary' parameters of that model. His methodological division is plausible, since the formal aspects of the standard model consist in certain symmetries, and certain symmetries (under translation, rotation, and reflection), were among Leibniz's favourite applications of the principle of sufficient reason. But most of the symmetries postulated by the standard model are 'internal' symmetries, and different internal symmetries are logically possible.What is the sufficient reason for nature's choice? Smolin's methodology, as he has worked it out so far, is not equipped to answer that question.

7. Wheeler

I turn now to John Archibald Wheeler, whose programme of 'law without law' antedated and inspired Smolin's programme and supplies, unfortunately only in outline, some elements missing in the latter. We cannot expect a definitive answer from Wheeler on the questions under consideration. He is extraordinarily exploratory, inventive, and self-critical, and he has criticized some of his earlier proposed answers – including 'superspace' and 'reprocessing of the universe'. What seems to me to be invariant over a period of five decades of exploration is a three-tiered methodological structure. The highest tier consists of one or more suggestive but somewhat cryptic aphorisms, such as 'genesis takes place through ownership' (1977, p. 29) and 'ask if the universe is not best conceived as a self-excited circuit' (1980, p. 362). The second tier consists in unpacking the aphorisms in order to achieve a mathematical theory, the main goal being the derivation of the principles of quantum mechanics. The third tier is required because the theory obtained at the second tier does not encompass all the fundamental laws of physics, but leaves room for expansion by evolution. The free play of natural selection in the presence of innumerable contingencies is supposed to generate the details of natural law. For example, the dimensionality of space-time is expected not to be fixed at the second tier in the methodology, but to be a consequence of the processes envisaged at the third tier.

In the past two decades Wheeler's first tier has emphasized *participation* of the observer in constituting the world. To prevent an infinite regress of explanation he says that 'no alternative is evident but loop, such a loop as this: "Physics gives rise to observation-participancy; observation-participancy gives rise to information; and information gives rise to physics"' (1990, p. 356). Wheeler claims to be following Niels Bohr's admonition that experimental conditions 'constitute an inherent element of the description of any phenomenon to which the term "physical reality" can be properly attached' (Bohr 1935, p. 699). But whereas Bohr uses his maxim to interpret the quantum principle, Wheeler has the ambition of deriving the quantum principle from his aphorisms (Wheeler 1979, p. 29). The derivation that he hopes for would be a central achievement of the second tier of his programme. He believes that it has been partially accomplished, at least to the point of proving that probabilities for microphysical transitions must be calculated by taking the absolute squares of complex amplitudes (1990, p. 359), but he considers the derivation of the totality of standard quantum theory to be still an open problem.

Since the ontological status of fundamental principles is the central concern of this paper, I shall focus attention on the first tier of Wheeler's programme. Supposing that the general principles of a participatory universe are purged of all obscurity, we wish to know the logical status of these principles. Are they *necessary*, and if so in what sense? Wheeler recognizes that his principles are close to the strong Anthropic Principle, which has been formulated as follows: 'The Universe must have those properties which allow life to develop within it at some stage in its history' (Barrow & Tipler 1986, p. 21). The crucial point is the character of this *must*. Wheeler hints at a belief that the necessity in question is logical necessity: 'That one can get so much from so little, almost everything from almost nothing, inspires hope that we will someday complete the mathematization of physics and derive everything from nothing, all law from no law' (1990, p. 358). I find this hope difficult to understand, because I see no logical inconsistency in the conception of a universe which is never aware of itself at any stage in its history. Furthermore, I am baffled that those who do find this conception to be inconsistent nevertheless accept (as Wheeler does) vast stretches of cosmic history in which awareness is not actual but latent. It may be that as a matter of ultimate fact the universe is so constituted as to ensure the emergence of awareness, but I see no sufficient reason for this fact without recourse to some kind of theology; and, as seen in our discussion of Leibniz, that move only postpones the encounter with a limit to the principle of sufficient reason.

8. Final remarks

To summarize, we have examined four thoughtful defences of the thesis that the fundamental laws of nature are products of evolution, and we have reached sceptical conclusions. Accordingly, our answer to questions about the ontological status of fundamental laws is 'uncertainty'. I suppose, however, that I shall not be permitted to conclude with this answer. There will be voices demanding, 'How shall we be uncertain, O Master?' Hence I shall go against my grain and make a few more remarks.

1. The programme of evolutionary explanation of the laws of nature is successful in many domains, surely in biology, probably in condensed matter physics, and possibly even in particle and space-time physics. How far the domain of its success extends is a scientific question the answer to which is inseparable from fundamental research in physics and cosmology.

2. However far the evolutionary explanation can be expressed, it will presuppose a theatre within which natural selection takes place, and this theatre must have some basic properties that are not susceptible to an evolutionary explanation.

3. From a strictly logical standpoint these basic properties are contingencies. They constitute the framework within which the entire cosmic drama takes place, however, and in this sense they are the necessities which constrain all natural laws. They are not necessities in the sense of intrinsic relations among the Forms, to use the language of Plato and Gödel, but rather necessities in the sense of being the uneliminable structural characteristics of the Receptacle (Plato, *Timaeus,* 49-51). Plato's notion of the Receptacle is cryptic, for the characteristics attributed to it are different in kind from those of any particular material substance, but that fact does not make the Receptacle completely characterless:

> It must be called always the same; for it never departs at all from its own character; since it is always receiving all things, and never in any way whatsoever takes on any character that is like any of the things that enter it: by nature it is there as a matrix for everything . . . (Ibid., 50B–50C)

4. Even if ontologically the most general principles have no sufficient reason, but simply are the basic facts about existence, epistemologically they appear to us as something quite different from the 'brute fact' of ordinary contingencies, such as the outcomes of indeterministic physical processes. A possible explanation for their epistemological singularity is the accommodation of the human intellect to the world in which we evolved. We are, after all, little gingerbread men baked in the great cosmic oven, and it would not be surprising that the basic properties of that oven have left their trace upon our natures.

5. The three tiers of Wheeler's programme are methodologically valuable even if they do not provide a satisfactory answer to our questions about the ontological status of fundamental natural laws. There is historical evidence that aphoristic guiding principles, which are the substance of Wheeler's first tier, are often very fruitful heuristically – think, for example of early versions of extremal principles, of the principle of nearby action, of conservation principles, of geometric invariance principles, and of the principle of equivalence. Why these should be fruitful generically is far from obvious. Perhaps the aphoristic mode of formulation enhances human abductive power in investigations at a high level of generality. Perhaps aphoristic formulation protects the scientist from being overwhelmed by an infinity of alternative laws, as can occur when inductive generalizations are proposed on the basis of a finite body of data. Wheeler's second tier – the derivation of mathematically precise theories from aphoristic principles – is evidently valuable when it succeeds, but even incomplete successes can be valuable, because perceptions of gaps in attempted derivations can prompt a reexamination

and tightening of the aphoristic principles of the first tier; Mach's principle, for example, has been subjected to reexamination in the light of general relativity theory (see, for example, Misner, Thorne, & Wheeler 1973, pp. 543–9). As to the third tier, the resort to natural selection in order to fill out the detailed structure of natural law is a humble acknowledgement of the pervasiveness of contingency in the world and the futility of aiming at a completely rational world picture. The bonus of this attitude of humility is a multitude of instances of attaining an understanding of the emergence of order out of disorder.

References

Barrow, J. & F. Tipler (1986). *The Anthropic Cosmological Principle* (Oxford: Oxford University Press).

Bohr, N. (1935). 'Can Quantum Mechanical Description of Physical Reality be Considered Complete?, *Physical Review* **48**, 696–702.

Conrad, J. (1900). *Lord Jim* (Blackwoods).

Kant, I. (1781). *Kritik der reinen Vernunft* (Riga: Johann Friedrich Hartknoch). Trans. N. Kemp Smith (London: Macmillan, 1929).

Leibniz, G. W. (1875–1890). *Die philosophischen Schriften*, ed. C. J. Gerhardt (Berlin) (collected from unpublished manuscripts and scattered publications of the seventeenth, eighteenth and nineteenth centuries).

Misner, C. W., K. S. Thorne & J. A. Wheeler (1973). *Gravitation* (San Francisco: Freeman).

Monod, J. (1972). *Chance and Necessity* (New York: Random House).

Peirce, C. S. (1934). *Collected Papers*, vol. 5, ed. C. Hartshorne and P. Weiss (Cambridge, MA: Harvard University Press).

(1935). *Collected Papers*, vol. 6, ed. C. Hartshorne & P. Weiss (Cambridge, MA: Harvard University Press).

Plato, *Timaeus*. Trans. with commentary F. M. Cornford, in Cornford, *Plato's Cosmology* (1937) (London: Routledge and Kegan Paul).

Redhead, M. (1995). *From Physics to Metaphysics* (Cambridge: Cambridge University Press).

Russell, B. (1900). *Philosophy of Leibniz* (London: George Allen & Unwin).

Shimony, A. (1989). 'The Non-Existence of a Principle of Natural Selection', *Biology and Philosophy* **4**, 255–73; reprinted in *Search for a Naturalistic World View*, vol. 2 (Cambridge: Cambridge University Press), pp. 228–52.

Smolin, L. (1997). *The Life of the Cosmos* (Oxford: Oxford University Press).

Wang, H. (1986). *Beyond Analytic Philosophy* (Cambridge, MA: MIT Press).

Wheeler, J. A. (1977). 'Genesis and Ownership', in R. Butts and J. Hintikka (eds.) *Foundational Problems in the Special Sciences* (Dordrecht: Reidel), pp. 3–33.

(1980). 'Beyond the Black Hole', in H. Woolf (ed.), *Some Strangeness in the Proportion: A Centennial Symposium to Celebrate the Achievements of Albert Einstein*, (Reading, MA: Addison-Wesley), pp. 341–75.

(1990). 'Information, Physics, Quantum: The Search for Links', in S. Kobayashi, H. Ezawa, Y. Murayama and S. Nomura (eds.) *Proceedings of the Third International Symposium: Foundations of Quantum Mechanics in the Light of New Technology* (Tokyo: The Physical Society of Japan), pp. 354–68.

Whitehead, A. N. (1929). *Process and Reality* (London: Macmillan).

(1933). *Adventures of Ideas* (London: Macmillan).

Bibliography of the writings of Michael Redhead

Books

Incompleteness, Nonlocality and Realism: A Prolegomenon to the Philosophy of Quantum Mechanics, pp. viii + 191, Oxford University Press, 1987
2nd revised impression (paperback), 1989
From Physics to Metaphysics, pp. xiii + 92, Cambridge University Press, 1995
paperback edition, 1996

Research articles

1 'Radiative Corrections to the Scattering of Electrons', *Proceedings of the Royal Society A* **220**, 219–239, 1953
2 'The Production of Bremsstrahlung in Electron–Electron Collisions', *Proceedings of the Physical Society A* **66**, 196–7, 1953
3 'Radiative Corrections to Coincidence Experiments in High Energy Electron–Electron and Positron–Electron Scattering', *Journal of Physics A* **5** 431–43, 1972
4 'On Neyman's Paradox and the Theory of Statistical Tests', *The British Journal for the Philosophy of Science* **25** 265–71, 1974
5 'Symmetry in Intertheory Relations', *Synthese* **32** 77–112, 1975
6 'Ad Hocness and the Appraisal of Theories', *The British Journal for the Philosophy of Science* **29** 355–61, 1978
7 'Models in Physics', *The British Journal for the Philosophy of Science* **31** 145–63, 1980
8 'Some Philosophical Aspects of Particle Physics', *Studies in History and Philosophy of Science* **11**, 279–304, 1980
9 'A Bayesian Reconstruction of the Methodology of Scientific Research Programmes', *Studies in History and Philosophy of Science* **11**, 341–7, 1980
10 'Experimental Tests of the Sum Rule', *Philosophy of Science* **48**, 50–64, 1981
11 'A Critique of the Disturbance Theory of Indeterminancy in Quantum Mechanics' (joint paper with H. R. Brown), *Foundations of Physics* **11**, 1–20, 1981
12 'Relativity, Causality and the Einstein–Podolsky–Rosen Paradox: Nonlocality and Peaceful Coexistence', in R. Swinburne (ed.), *Space, Time and Causality*, pp. 151–89 (Reidel, 1983)
13 'Nonlocality and the Kochen–Specker Paradox' (joint paper with P. Heywood), *Foundations of Physics* **13**, 481–99, 1983
14 'Quantum Field Theory of Philosophers', in P. D. Asquith and T. Nickles (eds), *Proceedings of the 1982 Biennial Meeting of the Philosophy of Science Association*, Vol. 2, pp. 57–99, 1983

15 'On the Impossibility of Inductive Probability', *The British Journal for the Philosophy of Science* **36**, 185–91, 1985

16 'Novelty and Confirmation', *The British Journal for the Philosophy of Science* **37**, 115–18, 1986

17 'Ontological Economy and Grand Unified Gauge Theories' (joint paper with J. S. Steigerwald), *Philosophy of Science* **53**, 280–1, 1986

18 'Relativity and Quantum Mechanics: Conflict or Peaceful Coexistence', *Annals of the New York Academy of Science* **480**, 14–20, 1986

19 'Whither Complementarity?', in N. Rescher (ed.), *Scientific Inquiry in Philosophical Perspective*, pp. 169–82 (University Press of America, 1987)

20 'A Philosopher looks at Quantum Field Theory', in H. R. Brown and R. Harré (eds), *Philosophical Foundations of Quantum Field Theory*, pp. 9–23 (Oxford University Press, 1988)

21 'Do the Bell Inequalities Require the Existence of Joint Probabilities?' (joint paper with G. Svetlichny, H. R. Brown and J. Butterfield), *Philosophy of Science* **55**, pp. 387–401, 1988

22 'The Compatibility of CP Violating Systems with Statistical Locality' (joint paper with R. K. Clifton), *Physics Letters A* **126**, pp. 295–9, 1988

23 'Quantum Physics and the Identity of Indiscernibles' (joint paper with S. French), *The British Journal for the Philosophy of Science* **39**, 233–46, 1988

24 'Nonfactorizability, Stochastic Causality and Passion-at-a-Distance', in J. Cushing and E. McMullin (eds.), *Philosophical Consequences of Quantum Theory*, pp. 145–53 (University of Notre Dame Press, 1989)

25 'Physics for Pedestrians', Inaugural Lecture (Cambridge University Press, 1989)

26 'Gibbs Paradox and Non-Uniform Convergence' (joint paper with K. G. Denbigh), *Synthese* **81**, 283–313, 1989

27 'The Nature of Reality', *The British Journal for the Philosophy of Science* **40**, 429–441, 1989

28 'Undressing Baby Bell', in R. Bhaskar (ed.), *Harré and His Critics*, pp. 122–8, (Oxford: Blackwell, 1990)

29 'Non-local Influences and Possible Worlds – A Stapp in the Wrong Direction' (joint paper with R. K. Clifton and J. Butterfield), *The British Journal for the Philosophy of Science* **41**, 5–58, 1990

30 'Explanation' in D. Knowles (ed.), *Explanation and its Limits*, pp. 135–154 (Cambridge: Cambridge University Press, 1990)

31 'Nonlocality and Quantum Mechanics', *Proceedings of the Aristotelian Society*, Supplementary Volume **65**, 119–40, 1991

32 'Particles, Particle Lables, and Quanta: The Toll of Unacknowledged Metaphysics' (joint paper with P. Teller), *Foundations of Physics* **21**, 43–62, 1991

33 'Generalisation of the Greenberger–Horne–Zeilinger Algebraic Proof of Nonlocality' (joint paper with R. K. Clifton and J. Butterfield), *Foundations of Physics* **21**, 149–84, 1991

34 'A Second Look at a Recent Algebraic Proof of Nonlocality' (joint paper with R. K. Clifton and J. Butterfield), *Foundations of Physics Letters* **4**, 395–403, 1991

35 'The Breakdown of Quantum Non-Locality in the Classical Limit' (joint paper with C. Pagonis and R. K. Clifton), *Physics Letters A* **155**, pp. 441–4, 1991

36 'Particle Labels and the Theory of Indistinguishable Particles in Quantum Mechanics' (joint paper with P. Teller), *The British Journal for the Philosophy of Science* **43**, 201–8, 1992

37 'Propensities, Correlations and Metaphysics', *Foundations of Physics* **22**, 381–94, 1992

38 'Is the End of Physics in Sight?', in S. French and H. Kamminga (eds.), *Correspondence, Invariance and Heuristics*, pp. 327–41 (Dordrecht: Kluwer, 1993)

39 'The Conventionality of Simultaneity', in J. Earman et al. (eds.), *Philosophical Problems of the Internal and External World: Essays Concerning the Philosophy of Adolf Grünbaum*, pp. 103–28 (University of Pittsburgh Press, 1994)

40 'Logic, Quanta and the Two-Slit Experiment', in P. Clark and R. Hale (eds.), *Reading Putnam*, pp. 161–75 (Oxford: Blackwell, 1994)

41 'Human Cognition and the Sciences' (joint paper with A. Musgrave), in *Wass Ist der Mensch? Menschenbilder im Wandel*, pp. 571–582, Europäisches Forum (Alpbach Ibera Verlag, 1994)

42 'More Ado About Nothing', *Foundations of Physics* **25**, 123–37, 1995

43 'Popper and the Quantum Theory', in A. O'Hear (ed.), *Karl Popper: Philosophy and Problems*, pp. 163–76 (Cambridge: Cambridge University Press, 1995)

44 'The Vacuum in Relativistic Quantum Field Theory', in D. Hull, M. Forbes & R. M. Burian (eds.), *Proceedings of the 1994 Biennial Meeting of Philosophy of Science Association*, Vol. 2, pp. 77–87, 1995

45 'EPR, Relativity and the GHZ Experiment' (joint paper with C. Pagonis and P. La Rivière), in R. Clifton (ed.), *Perspectives on Quantum Reality: Non-Relativistic, Relativistic, and Field-Theoretic*, pp. 43–55 (Dordrecht: Kluwer, 1996)

46 'The Twin Paradox and the Conventionality of Simultaneity' (joint paper with T. Debs), *The American Journal of Physics* **64**, 384–92, 1996

47 'A Value Rule for Non-Maximal Observables' (joint paper with A. Stocks), *Foundations of Physics Letters* **9**, 109–19, 1996

48 'The Relativistic EPR Argument' (joint paper with P. La Rivière), in R. Cohen, M. Horne & J. Stachel (eds.), *Potentiality, Entanglement, and Passion-at-a-Distance: Quantum Mechanical Studies for Abner Shimony*, Vol. 2, pp. 207–215 (Dordrecht: Kluwer, 1997)

49 'Should We Believe in Quarks and QCD?', in L. Hoddesdon, L. Brown, M. Riordan & M. Dresden (eds.), *The Rise of the Standard Model*, pp. 637–644, (Cambridge: Cambridge University Press, 1997)

50 'The Hahn Spin-Echo Experiment and the Second Law of Thermodynamics' (joint paper with T. N. Ridderbos), *Foundations of Physics*, **28**, 1237–70

51 'Superentangled States' (joint paper with R. Clifton, David Feldman, Hans Halvorson and Alexander Wilce), *Physical Review A* **58**, 135–45, 1998

52 'Unified Treatment of EPR and Bell Arguments in Algebraic Quantum Field Theory' (joint paper with F. Wagner), *Foundations of Physics Letters* **11**, 111–25, 1998

53 'Quantum Field Theory and the Philosopher' in T. Y. Cao (ed.), *Conceptual Foundations of Quantum Field Theory* (Cambridge: Cambridge University Press, 1999), pp. 34–40.

Review and encyclopaedia articles

1 'Orthodoxy in Quantum Mechanics' (Critical notice of E. Scheibe, *The Logical Analysis of Quantum Mechanics*), *The British Journal for the Philosophy of Science* **25**, 353–8, 1974

2 'Wave-Particle Duality' (Critical notice of M. Audit, *The Interpretation of Quantum Mechanics*), *The British Journal for the Philosophy of Science* **28**, 65–74, 1977

3 Article on Einstein in J. Wintle (ed.), *Makers of Modern Culture*, pp. 146–9 (London: Routledge and Kegan Paul, 1981)

4 Article on Boole in J. Wintle (ed.), *Makers of Nineteenth Century Culture*, pp. 62–3 (London: Routledge and Kegan Paul, 1982)

5 Article on Duhem in J. Wintle (ed.), *Makers of Nineteenth Century Culture*, pp. 181–3, (London: Routledge and Kegan Paul, 1982)

6 'Unification in Science' (Critical notice of C. F. von Weizsäcker, *The Unity of Nature*), *The British Journal for the Philosophy of Science* **35**, 274–9, 1984

7 Article on 'Quantum Theory' in G. N. Cantor, J. R. R. Christie, M. J. S. Hodge and R. C. Olby (eds.), *Companion to the History of Modern Science*, pp. 458–78 (London: Routledge, 1989)

8 'No Boundaries: Hawking's Universe', *The Cambridge Review* **113**, pp. 8–10, March 1992

9 Article on the Philosophical Consequences of Relativity Theory, in the *Routledge Encyclopedia of Philosophy*, ed. E. Craig, vol. 8, pp. 191–200, 1998

Book reviews

1 Review of Max Jammer, *The Philosophy of Quantum Mechanics, Times Higher Education Supplement*, 15 August 1975

2 Review of C. Hooker (ed.), *The Logico-Algebraic Approach to Quantum Mechanics, Annals of Science* **84**, 631–2, 1977

3 Review of Rudolf Carnap, *Two Essays on Entropy, The Philosophical Quarterly* **29**, 364–6, 1979

4 Review of D. Hull and O. Prewett (eds.), *The Nature of the Physical Universe, The British Journal for the Philosophy of Science* **31**, 404–8, 1980

5 Review of J. Ziman, *Reliable Knowledge, The British Journal for the Philosophy of Science* **32**, 311–14, 1981

6 Review of R. Peierls, *Surprises in Theoretical Research Physics, The British Journal for the Philosophy of Science* **32**, 309–11, 1981

7 Review of A. Pais, *Subtle is the Lord: The Science and Life of Albert Einstein, The British Journal for the History of Science* **77**, 226–7, 1984

8 Review of A. van der Merwe (ed.), *Old and New Questions in Physics, Cosmology, Philosophy and Theoretical Biology: Essays in Honour of Wolfgang Yourgrau, Endeavour* **8**, 4, 1984

9 Review of J. A. Wheeler and W. H. Zurek (eds.), *Quantum Theory and Measurement, Endeavour* **8**, 152, 1984

10 Review of N. Cartwright, *How the Laws of Physics Lie*, *The Philosophical Quarterly* **34**, pp. 513–14, 1984

11 Review of R. Torretti, *Relativity and Geometry*, *The British Journal for the Philosophy of Science* **36**, 100–4, 1985

12 Review of J. Hendry, *The Creation of Quantum Mechanics and the Bohr–Pauli Dialogue*, *Endeavour* **9**, 60, 1985

13 Review of P. C. W. Davies and J. R. Brown (eds.), *The Ghost in the Atom: A Discussion of the Mysteries of Quantum Physics*, *Nature* **324**, 420–1, 1986

14 Review of J. Lucas, *Space, Time and Causality*, *The Philosophical Quarterly* **36**, 453–7, 1986

15 Review of G. Holton, *The Advancement of Science and its Burdens*, *Endeavour* **11**, 216, 1987

16 Review of A. Louizou, *The Reality of Time*, *Philosophical Books* **29**, 118–9, 1988

17 Review of B. N. Kursonoglu and E. P. Wigner (eds.), *Reminiscences About a Great Physicist: Paul Adrien Dirac*, *Endeavour*, **12**, 97, 1988

18 Review of D. Murdoch, *Niels Bohr's Philosophy of Physics*, and J. S. Bell, *Speakable and Unspeakable in Quantum Mechanics*, *Nature* **331** 667, 1988

19 Review of R. J. Russell, W. B. Stoeger & G. V. Coyne, *Physics and Theology, a Common Quest for Understanding*, *Annals of Science* **46**, 424–5, 1989

20 Review of P. Galison, *How Experiments End*, *Synthese* **82**, 157–62, 1990

21 Review of B. d'Espagnat, *Reality and the Physicist*, *The Philosophical Quarterly* **40**, 257–8, 1990

22 Review of J. Polkinghorne, *The Rochester Roundabout*, *Annals of Science* **48**, 101–2, 1991

23 Review of J. Barrow, *Theories of Everything*, *Times Literary Supplement*, 26 July 1991

24 Review of J. Earman, *World Enough and Space-Time*, *Philosophy of Science* **59**, 718–22, 1992

25 Review of A. Pais, *Niels Bohr's Times*, *Notes and Records of the Royal Society*, **47**, 152–4, 1993

26 Review of J. Earman, *Bayes or Bust*, *The Times Higher Education Supplement*, 19 March 1993

27 Review of D. Lindley, *The End of Physics: The Myth of a Unified Theory*, *The Times Higher Education Supplement*, 30 September 1994

28 Review of A. C. Crombie, *Styles of Thinking in the European Tradition*, to appear in *The Times Higher Education Supplement*, 30 June 1995

29 Review of S. Petruccioli, *Atoms, Metaphors and Paradoxes*, *Contemporary Physics* **37**, 77, 1996

30 Review of W. L. Craig and Q. Smith, *Theism, Aetheism, and the Big Bang Cosmology*, *The British Journal for the Philosophy of Science* **47**, 133–6, 1996

31 Review of T. Maudlin, *Quantum Non-Locality and Relativity*, *The Philosophical Quarterly* **47**, 118–20, 1997

32 Review of A. Shimony, *Search for a Naturalistic World View*, *Synthese* **110**, 335–42, 1997

33 Review of R. M. Hazen and M. Singer, *Why Aren't Black Holes Black?: The Unanswered Questions at the Frontiers of Science*, *The New York Times*, 28 September 1997

34 Review of D. Deutsch, and L. Smolin, *The Life of the Cosmos*, *The Times Literary Supplement*, 2 January 1998
35 Review of S. Auyang, *How is Quantum Field Theory Possible?*, *The British Journal for the Philosophy of Science* **44**, 499–507, 1998
36 Review of Jim Cushing, *Philosophical Concepts in Physics: The Historical Relation Between Philosophy and Scientific Theories*, *Physics World* 43–4, June 1998

Index of names

Index of subjects

Lightning Source UK Ltd.
Milton Keynes UK
UKHW012149050819
347466UK00001B/19/P

9 780521 154475